Biochemistry

Dushyant Kumar Sharma

Alpha Science International Ltd.
Oxford, U.K.

Biochemistry

280 pgs. | 84 figs. | 17 tbls.

Dushyant Kumar Sharma
Department of Zoology
Government P.G. College
Morena, India

ALPHA SCIENCE INTERNATIONAL LTD.

7200 The Quorum, Oxford Business Park North
Garsington Road, Oxford OX4 2JZ, U.K.

www.alphasci.com

ISBN 978-1-84265-510-8

Printed in India

Dedicated

to

my father

Late Shri L.P. Sharma

Preface

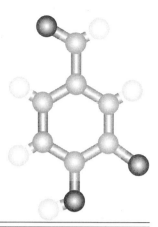

Biochemistry is concerned with the chemistry of life. It uses the principles and concepts of chemistry for elucidation of biological processes. It includes not only the study of chemical substances occurring in the living system but also the physiology of various vital processes of living beings. In recent years, biochemistry has emerged as a distinct discipline with wide applications in various fields of science. It is no more an isolated subject rather it is closely related to different branches of science.

Though a large number of books are available on the subject, there is a need for a book which could explain complicated concepts in biochemistry in a simplified manner. The present book is a sincere and humble effort in this direction. The book has been written keeping the students' need and interest in mind.

The book has been organized into four parts. The first part deals with the study of structure and functions of various biomolecules. The second part describes the metabolism of carbohydrates, lipids, proteins, nucleic acids and inborn metabolic errors. The third part briefly discusses the gene and gene expression. The fourth part focuses on the techniques commonly used in biochemistry.

I wish to express my thanks to Mr. N.K.Mehra, Narosa Publishing House, New Delhi and his wonderful team for bringing out this book in this impressive form.

I am grateful to all my colleagues and friends who have contributed, directly or indirectly, in the completion of this book

I would like to convey my thanks to my wife and children for their moral support, cooperation and encouragement.

Though every effort has been made to make the book up-to-date and free from mistakes but biochemistry is subject which is so vast and ever-expanding that there may still be some deficiencies. I would appreciate constructive criticism and suggestions from the readers.

Finally, I present this book with the hope that it will be able to develop a keen interest among the students in biochemistry.

Dushyant Kumar Sharma

Contents

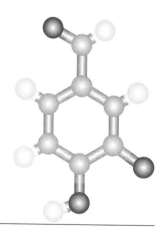

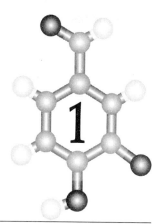

Introduction

1.1 INTRODUCTION

To understand the mystery of life has always been man's biggest desire. Biochemistry, the study of chemistry of life has always tried to quench the thirst of man in this direction. Biochemistry is a branch of science which deals with the study of complex chemical reactions and chemical structures which give rise to life and life processes. It is the application of chemistry to study the biological processes at the cellular and molecular level. In biochemistry, we study the structures and functions of biomolecules such as proteins, carbohydrates, lipids and nucleic acids which constitute cell-the basic unit of organisms. Cell is the structural and functional unit of an organism. A cell is nothing but an assemblage of various chemicals in a particular ratio.

If we go back to the origin of life, it is quite clear that simple molecules combined to form complex molecules which ultimately led to the origin of life.

Biochemistry is an interdisciplinary subject which is closely related to other fields of biology such as cell biology, genetics, molecular biology, microbiology, immunology, biotechnology and biophysics. There has never been a significant difference between biochemistry and in these disciplines in terms of content and techniques. Researchers in biochemistry use techniques and ideas from different disciplines of science. With the involvement of methods of physics, chemistry and biology, biochemistry aims not only to find out the structure of complex molecules present in biological materials but the ways these molecules interact to form cell, tissues and the whole organism.

1.2 BRIEF HISTORY OF BIOCHEMISTRY

Biochemistry is relatively a young branch of science which has recently emerged as a distinct discipline of science.

Biochemistry started as an off shoot of organic chemistry. The term 'biochemistry' was coined by *Carl Neuberg* in 1903. The history of biochemistry begins with the isolation of urea by *Rouelle* from urine in 1773. In 1780, *Scheele* isolated several organic compounds such as citric acid, malic acid, glycerol and uric acid from natural resources. In 1789, *Antoine Lavoisier* developed the

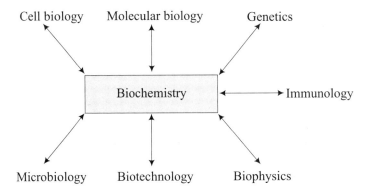

Fig. 1   Biochemistry : An inter disciplinary science

concept of oxidation and respiration. In 1828, *Friedrich Wohler* synthesized urea from inorganic precursors. It was for the first time that an organic compound was created artificially, marking the real beginning of biochemistry.

In 1878, *F.W. Kuhne* coined the term 'enzyme'. The breakthrough in enzyme chemistry was made by *Edward Buchner* who demonstrated alcoholic fermentation outside of a cell in 1896. Significant advancements have been made in biochemistry in the twentieth century. In 1926 *J.Sumner* isolated the enzyme urease for the first time. *Stanley* isolated the first crystallized virus TMV (tobacco mosaic virus). In 1937, *Krebs* discovered Krebs Cycle (citric acid cycle). In 1953 *Watson* and *Crick* proposed the double helical model of DNA for which they were awarded Nobel Prize. In 1958 *Beadle and Tatum* received Nobel Prize for their work in fungi demonstrating "one gene produces one enzyme". *Milstein*, in 1975, developed hybridoma cells and monoclonal antibodies. In 1980 *Paul Berg* received Nobel Prize for his work on recombinant DNA molecules. Nobel Prize for 1986 was awarded to *Rita-Levi Montalcini* and *Stanley Cohen* for their discovery of a neural growth factor. *Susuma Tonegawa* studied the biochemistry of immune defence mechanism and generation of antibody diversity for which he was awarded Nobel Prize in 1987. In 1988 *Johann Deisenhofer*, *Robert Huber* and *Hartmut Michel* determined the three dimensional structure of photosynthetic reaction center, the major photochemical energy processing machinary in plants and bacterial photosynthesis.

With advancement in technology and analytical approach, biochemistry has grown tremendously in the last two decades. Recently, *Andrew Z. Fire* and *Craig C. Mellow* were awarded Nobel Prize for discovering the role of RNA interference in the silencing of gene expression, in 2006. Some historical events have been summerized in Table 1.

1.3   SCOPE OF BIOCHEMISTRY

Biochemistry is the foundation for understanding various biological processes.The knowledge and methods of biochemistry can be applied to different fields like medicine, agriculture, biotechnology and other related fields to reveal the complexity of the organisms. Biochemistry provides basis for the practical advances in medicine, veterinary, agriculture and biotechnology.

Table 1   Some historical events in biochemistry

Year	Scientist	Event
1780	Scheele	Isolation of organic compounds from natural resources.
1828	F. Wohler	Synthesis of urea
1878	F.W. Kuhne	Coined the term enzyme
1898	Fischer	Lock and key concept of enzyme action
1903	Carl Neuberg	The term biochemistry
1914	Kendell	Isolation of thyroxine
1926	Sumner	Isolation of enzyme urease
1935	Stanley	Isolation of tobacco mosaic virus (TMV)
1937	Krebs	Citric Acid cycle
1948	Calvin and Bensen	Calvin cycle in photosynthesis
1953	Watson and Crick	Double helical model of DNA
1975	Mitstein	Hybridoma cells
1986	Rita-Levi Montalcini Stanley Cohen	Neural growth factor
2006	A.Z. Fire, C.C. Mellow	Role of RNA interference in gene expression

Advances made in biochemistry have made significant contribution towards elucidation and understanding the mystery of life. With advancement in technology biochemistry has broadened its area. Various subdisciplines of biochemistry such as clinical biochemistry, physical biochemistry, neurochemistry and immunochemistry have emerged recently.

Clinical biochemistry is an important area of biochemistry which aims at the diagnosis and treatment of diseases. Any change in the normal physiology of an organism which leads to a disease causes the changes in the biochemistry of the body. For instance, in most of the diseases the changes appear in the chemistry of blood. The changes may be in plasma, serum or blood cells. There may be an increase or decrease in the concentration of blood components or appearance of some new substances in circulation. These alterations signal the onset of some disease. By studying the biochemistry of blood and other body fluids certain diseases can be diagnosed and treatment can be found out accordingly. Biochemistry is highly significant in pharmaceutical science. Synthesis of drugs, hormones, antibiotics etc. requires the principle and techniques of biochemistry.

In the field of agriculture, biochemistry focuses on the development and innovation of new techniques for crop cultivation, pest control management and crop storage management. New varieties of plants which are disease and drought resistant are being developed in collaboration with biotechnology. New improved varieties of plants and animals can be produced by applying the techniques of biochemistry in biotechnology. Biochemistry has contributed significantly in the development of biotechnology.

Biochemical methods are widely used in different industries. Use of enzymes in food, detergent and personal care products industries is not new.

Economic development and industrialization have led to one of the major problems threatening the very existence of mankind on the planet–pollution.

Pollution is increasing at an alarming speed. Biochemical methods for disposal of industrial and domestic wastes are the need of hour. Biochemistry could provide good solution in controlling pollution in a safe and effective way.

In addition to the above mentioned areas, biochemistry as a subject has lot of career prospects and job opportunities. There is a great demand of biochemistry in various research institutes and industries. In hospitals, clinical biochemists are engaged in pathological labs. Pharmaceutical and agriculture research based industries require biochemists in good number.

Thus, it can be concluded that biochemistry is the science which has solutions to all our problems we are facing today or we may face in future. Biochemistry has emerged as a discipline which has captivated the minds of scientific community with the aim of making our life more comfortable and fulfilling.

Part I

Biomolecules

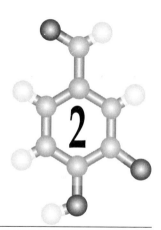

Physical Properties of Molecules

2.1 ACIDS AND BASES

There are different theories to explain acids and bases.

2.1.1 Arrhenius Concept of Acids and Bases

According to Arrhenius concept, an acid is a substance which can give hydrogen ion in aqueous solution, and a base is a substance which gives hydroxide ion in aqueous solution.

Acids:
$$HNO_{3(aq)} \rightleftharpoons H^+_{(aq)} + NO^-_{3(aq)}$$
$$HCl_{(aq)} \rightleftharpoons H^+_{(aq)} + Cl^-_{(aq)}$$
$$CH_3COOH_{(aq)} \rightleftharpoons CH_3COO^-_{(aq)} + H^+_{(aq)}$$

Bases:
$$NaOH_{(aq)} \rightleftharpoons Na^+_{(aq)} + OH^-_{(aq)}$$
$$NH_4OH_{(aq)} \rightleftharpoons NH_4^+_{(aq)} + OH^-_{(aq)}$$

Acids, such as HCl, HNO_3 which are almost completely ionized in aqueous solution are termed as *strong acids,* while acids such as CH_3COOH which are weakly ionized are termed as *weak acids.*

Similarly, bases which are completely ionized in aqueous solution such as NaOH are called as *strong bases* and the bases which are slightly ionized such as NH_4OH are called *weak bases.* These definitions of acids and bases may be applied only for those reactions which take place in aqueous solutions.

2.1.2 Bronsted-Lowry Concept

In 1923, Bronsted and Lowry proposed new definitions for acids and bases. According to them, an acid is a substance which can donate a proton and a base is a substance which can accept a proton.

$$\underset{\text{Base}}{\underline{A}} + H^+ \rightleftharpoons \underset{\text{Acid}}{\underline{A^+H}}$$

In the above reaction, A accepts a proton and thus behaves as a base, while A^+ H gives up a proton and acts as an acid.

$$\underset{\text{Acid}}{\underline{HCl}} \text{ (aq)} + \underset{\text{Base}}{\underline{H_2O}} \text{ (l)} \rightleftharpoons H_3O^-_{(aq)} + Cl^-_{(aq)}$$

When an acid loses a proton it becomes a base whereas a base after accepting a proton becomes an acid. Let us consider a reaction,

$$\underset{\text{Acid}}{H_2O \text{ (l)}} + \underset{\text{Base}}{NH_{3(aq)}} \rightleftharpoons \underset{\text{Acid}}{NH^+_{4 \text{ (aq)}}} + \underset{\text{Base}}{OH^- \text{(aq)}}$$

In this reaction, water donates a proton to ammonia (base) and acts as an acid. Similarly, in the reverse reaction, NH_4^+ ions donate a proton to the OH^- ions and act as acid. A base formed by the loss of proton by an acid is called *conjugate base* of the acid where as acid formed by accepting a proton by the base is called *conjugate acid* of the base. Here, OH^- is the conjugate base of H_2O and NH_4^+ is conjugate acid of NH_3. These acid-base pairs (H_2O/OH^-) and (NH_4^+/NH_3) are called as *conjugate acid-base pairs*.

There are certain substances which can act as acids as well as bases. Such substances are called *amphoteric substances*. For example,

(i) $CH_3COOH_{(aq)} + H_2O_{(l)} \rightleftharpoons H_3O^+_{(aq)} + CH_3COO^-_{(aq)}$

(ii) $H_2O_{(l)} + CO_3^{2-}_{(aq)} \rightleftharpoons HCO_{3(aq)}^- + OH^-_{(aq)}$

In the first reaction, H_2O behaves as a base (accepts proton) and in the second reaction, H_2O behaves as an acid (donates proton).

2.1.3 Lewis Concept

G.N. Lewis proposed more general definitions of acids and bases. According to this, 'an acid is a substance which can accept a pair of electrons and a base is a substance which can donate a pair of electrons'.

According to this concept acid-base reactions involve donation of electron pair by a base to an acid to form a coordinate bond. Lewis bases can be neutral molecules such as $\ddot{N}H_3$, $CH_3-\ddot{O}H$, $H_2\ddot{O}$ etc. having one or more unshared pairs of electrons or anions such as CN^-, OH^-, or Cl^-. Lewis acids are the species having vacant orbitals in the valence shell of one of its atoms.

Lewis Base   Lewis Acid

$$H\!:\!\overset{..}{\underset{H}{O}}\!:\quad + \quad H^+ \quad \longrightarrow \quad \left[H\!:\!\overset{..}{\underset{H}{O}}\!:\!H\right]^+$$

Lewis Base   Lewis Acid

$$Si\,F_4 \;+\; 2\;:\!\overset{..}{\underset{..}{F}}\!:\quad \longrightarrow \quad \left[Si\,F_6\right]^{2-}$$

Lewis concept of acids and bases is more general than Arrhenius and Bronsted concept.

2.2 DIFFUSION

When we put a few drops of blue ink in a glass of water, the ink disperses in the water and the colour of the water becomes blue. (Fig. 2.1) Similarly, when we spray a perfume, the perfume is smelled in all parts of the room. These are the examples of diffusion. Diffusion is the movement of molecules from higher concentration to lower concentration down a concentration gradient. Diffusion is a *passive movement* i.e. molecules move at their own—the process does not require energy. Diffusion is seen in liquids, gases and in solids. Diffusion is more rapid in gases than in liquids, and solids. In liquids the movement is more in comparison to solids, where the movement of molecules is slow and some times it is zero. Diffusion is a consequence of thermal motion of atoms, molecules and particles which results in the movement of particles from higher concentration to lower concentration. There are various factors which affect the rate of diffusion. Diffusion is directly proportional to temperature. The rate of diffusion increases with temperature. The temperature of a system is a measure of the average kinetic energy. A higher kinetic energy means a higher velocity. Thus, the speed of diffusion increases with temperature. Diffusion depends on the size and shape of the solute particles. Large particles diffuse more slowly than the small particles. Diffusion of a gas or solute also depends on its solubility in the medium. Viscosity of the medium also affects the rate of diffusion. The rate of diffusion is slow in more viscous solutions.

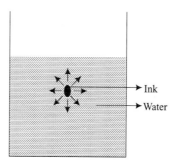

Fig. 2.1 Diffusion

2.2.1 Biological Significance

A number of processes in biological systems take place by diffusion. For example, diffusion is the main form of transport of water, ions and small molecules through cell membrane. Exchange of gases (O_2 and CO_2) between blood and lungs and between blood and tissues takes place by simple diffusion. Exchange of materials between RBCs and plasma occurs by diffusion. Absorption of certain food materials in intestine occurs by diffusion. During urine formation, renal absorption of urea is carried out by the process of diffusion.

2.3 OSMOSIS

Osmosis is the movement of water molecules from higher concentration to lower concentration through a semipermeable membrane. A semipermeable membrane allows the movement of water molecules but restricts the movement of solute particles. When two solutions are separated by a semipermeable membrane water molecules diffuse through the membrane from the side of low solute concentration (hypotonic) to the side of high solute concentration (hypertonic) till the concentrations of both sides of solutions become equal. For example, if two salt solutions, say 1% NaCl and 9% NaCl solutions are separated by a semipermeable membrane, water will move from 1% NaCl solution to 9% NaCl solution till the concentrations of both solutions become equal–equilibrium is reached (Fig. 2.2). At this stage a pressure develops which prevents further flow of water through the membrane. This is the *Osmotic pressure*. Osmotic pressure is the pressure required to prevent the flow of water in the concentrated solution from the dilute solution through a semipermeable membrane. Osmotic pressure of a solution depends on the concentration of solute. The osmotic pressure of a highly concentrated solution (hypertonic) will be more than that of a dilute (hypotonic) solution. Osmotic pressure of two isotonic solutions will be equal. Osmotic pressure is a 'colligative property' which means that it depends on the molar concentration of the solute and not upon the composition of solute.

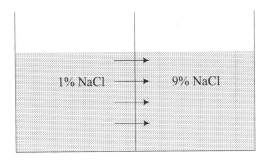

Fig. 2.2 Osmosis

2.3.1 Biological Significance

Osmosis is very important in biological systems. In plants, osmosis is important for many functions. Absorption of water by roots is by osmosis. In certain plants, such as herbaceous plants, osmotic pressure raises the turgor pressure in the cell wall which supports them to stand upright.

Movement of water, in and out of the cell is by osmosis. Cell membrane acts as a semipermeable membrane which allows movement of water and some small molecules and restricts the movement of large molecules. In animals, absorption of certain food materials is by osmosis.

Exchange of fluids among blood, intracellular fluids and cells is by osmosis. When an animal cell or RBC is placed in salt or sugar solutions of different concentrations, different changes are observed. For example, when a cell is placed in a hypotonic solution water will diffuse into the cell and the cell will swell and may burst. When a cell is placed in a hypertonic solution, water will diffuse out of the cell and the cell will shrivel. If the solution is isotonic, there will be no net movement across the cell membrane.

2.4 VISCOSITY

It is a common observation that all liquids do not flow at the same speed. Some liquids such as oil and honey flow at a slow speed, while others such as water flow rapidly. The difference in the flow of liquids is due to viscosity. Viscosity is the measure of fluid's resistance to flow. It is the measure of internal friction of a fluid. The greater the friction, the greater the amount of force required to cause the movement. A fluid with low viscosity flows easily while a fluid with large viscosity resists motion. Water has a low viscosity and flows easily while oil having high viscosity does not flow as easily as water.

The fundamental unit of viscosity is *poise*. A poise is the force in dynes, required to be applied to an area of 1 sq. cm between two parallel planes of 1 sq. cm in area and 1cm. apart, to produce a difference in stream velocity between the liquid planes of 1 cm/sec. Viscosity is affected by various factors such as temperature, density and size and shape of solute. The viscosity of a liquid varies with temperature. With the rise in temperature, viscosity decreases. Viscosity is directly proportional to density. Viscosity also varies with size and symmetry of solutes. Liquids with large molecules are more viscous.

The term 'fluidity' is opposite to viscosity and expresses the tendency of a liquid to flow.

2.5 pH

Hydrogen ion concentration of a solution is the number of gram ions of hydrogen in one litre of a solution. The acidity or alkalinity of a solution is expressed in terms of hydrogen ion concentration.

In case of pure water, which is slightly ionized into $[H^+]$ and $[OH^-]$, the concentration of $[H^+]$ and $[OH^-]$ is equal and is 10^{-7} per litre. The product of concentration of $[H^+]$ and $[OH^-]$ in pure water is 1.0×10^{-14} at 25°C and is expressed as:

$$K_w = [H^+] \times [OH^-]$$
$$= 1.0 \times 10^{-14}$$

Since, $\qquad [H^+] = [OH^-]$

thus, $\qquad K_w = [H^+] [H^+]$
$$= [H^+]^2$$
$$= 1.0 \times 10^{-14}$$

or, $\qquad [H^+] = 1 \times 10^{-7}$

In different solutions, the concentrations of $[H^+]$ and $[OH^-]$ are different. But at a particular temperature, the product of $[H^+]$ and $[OH^-]$ remains constant.

In a solution, if concentration of $[H^+]$ is more than $[OH^-]$, it will be an *acid*. If concentration of $[OH^-]$ is more than $[H^+]$, the solution will be a *base*. In case of neutral solutions, conentration of $[H^+] = [OH^-]$.

Thus, if

(i) $[H^+] = [OH^-] = 10^{-7}$ = Neutral solution.

(ii) $[H^+] > [OH^-]$
      $> 10^{-7} =$ Acidic solution.

(iii) $[H^+] < [OH^-]$
      $< 10^{-7} =$ Alkaline solution.

To express the hydrogen ion concentration, *Soren Sorensen* in 1909 introduced a scale, which is known as pH scale. In the expression pH, p stands for power and H stands for hydrogen ion. pH of a solution is "the negative logarithm of its hydrogen ion concentration".

$$pH = - \log [H^+]$$

$$= \log \frac{1}{[H^+]}$$

For pure water,

$$= \log \frac{1}{1 \times 10^{-7}}$$

$$= \log [1 \times 10^7]$$

$$= \log 1.0 + \log 10^7$$

$$= 0 + 7$$

$$= 7$$

pH of pure water is 7 (neutral). In a solution if concentration of $[H^+]$ is 10^{-5}, its pH will be 5, if the concentration of $[H^+]$ is 10^{-10}, the pH will be 10.

When pH is less than 7 i.e. 0-7, the solution is acidic and when pH is more than 7 i.e. 7-14, the solution is alkaline in nature.

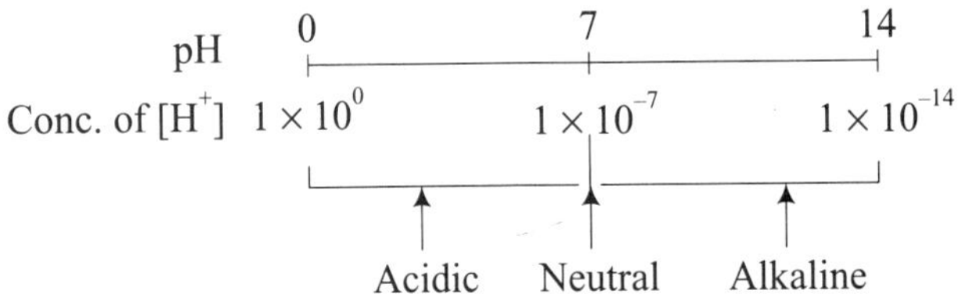

pH   0   7   14
Conc. of $[H^+]$ 1×10^0 1×10^{-7} 1×10^{-14}

                  Acidic   Neutral   Alkaline

A lower pH value indicates increasing strength of acidity and a higher pH indicates increasing strength of alkalinity.

2.5.1   Determination of pH

The methods which are generally used for determination of pH are:
  (a) pH-indicator method
  (b) Potentiometric method

(a) *pH indicator method:*   pH of a solution can be determined by addition of a pH indicator into the solution. A number of organic substances which change their colour with a slight change in pH are used as pH indicators (Table 1). For example, phenolphthalein which is a commonly used indicator, is colourless below a pH of 8.3 and distinctly pink above 10. Between 8.3 and 10 it exhibits different shades of pink with respect to pH. By using this indicator, pH of a solution can be measured within a range of pH. pH paper indicator is also widely used for measuring pH. pH paper is usually a strip of paper that has been soaked in an indicator solution. It changes its colour with the change in the pH of the solution.

Table 2.1   Some commonly used pH indicators

Indicator	pH Range	Colour Change
1. Methyl blue	2.9-4.0	red to yellow
2. Bromophenol blue	3.0-4.6	yellow to blue
3. Methyl red	4.4-6.0	red to yellow
4. Methyl orange	3.1-4.4	red to yellow
5. Phenol red	6.8-8.4	yellow to red
6. Acyl blue	12.0-13.6	red to blue
7. Phenolphthalein	8.3-10.0	colourless to pink

(b) *Potentiometric method:*   In this method, an electrode or pH meter is used to determine the pH of a solution. Different types of electrodes-glass electrode, hydrogen electrode, quinhydrone electrode, antimony-antimony oxide electrode and ion-selective electrode are used for different solutions.

2.5.2   Significance of pH

pH has great significance in biological systems. In the body, metabolic reactions take place at an optimal pH. The enzymes act at an optimal pH. Any deviation from this pH affects the rate of metabolic reactions. In body different body fluids have specific pH and the body follows various mechanisms to maintain pH.

2.6   BUFFER SOLUTION

Generally, when an acid or base is added to a solution it changes its pH. But there are certain solutions which resist any change in pH when a little amount of acid or base is added to them. Such solutions are called *buffer solutions.*

"A buffer solution is the solution which can resist the change in pH on addition of small amount of acid or base. This ability of a buffer solution to resist change in pH on addition of an acid or base is called *buffer action*."

The buffer may be an *acid buffer* or *basic buffer*. Acid buffers contain a weak acid and its salt. e.g. acetic acid and sodium acetate. Basic buffers contain a weak base and its salt. e.g. ammonium hydroxide and ammonium chloride.

2.6.1 Acid Buffers

A weak acid and its salt act as acid buffer. Acetic acid and sodium acetate act as acid buffer. Acetic acid, being a weak acid is weakly dissociated and sodium acetate being a salt ionizes completely to form CH_3COO^- and Na^+.

$$CH_3COOH \rightleftharpoons CH_3COO^- + H^+$$
$$CH_3COONa \rightleftharpoons CH_3COO^- + Na^+$$

The ionization of acetic acid is further suppressed by the acetate ions from sodium acetate (common ion effect).

When a few drops of an acid (e.g. HCl) are added to it, the H^+ ions combine with CH_3COO^- ions to form weakly dissociable CH_3COOH. Thus, there is no rise in H^+ ion concentration and pH remains unchanged. Cl^- ions combine with Na^+ ions to from NaCl.

$$HCl \rightleftharpoons H^+ + Cl^-$$
$$H^+ + CH_3COO^- \rightleftharpoons CH_3COOH$$
$$Na^+ + Cl^- \rightleftharpoons NaCl$$

Similarly, when a few drops of a base (e.g. NaOH) are added to buffer solution, the OH^- ions combine with H^+ to form water and Na^+ ions combine with CH_3COO^- ions to form CH_3COO Na.

$$NaOH \rightleftharpoons Na^+ + OH^-$$
$$H^+ + OH^- \rightleftharpoons H_2O$$
$$CH_3COO^- + Na^+ \rightleftharpoons CH_3COONa.$$

Thus by addition of a few drops of NaOH, no change in pH of the solution is observed.

2.6.2 Basic Buffers

NH_4OH and NH_4Cl act as basic buffer. NH_4OH being a weak base is slightly ionized, while NH_4Cl is completely ionized to NH_4^+ and Cl^-

$$NH_4OH \rightleftharpoons NH_4^+ + OH^-$$
$$NH_4Cl \rightleftharpoons NH_4^+ + Cl^-$$

NH_4^+ ions further suppress the ionization of NH_4OH which is already a weakly ionizable substance. When a few drops of a base (e.g. NaOH) are added to it, the OH^- ions from NaOH combine with NH_4^+ ions to form weakly ionizable NH_4OH and thus the concentration of OH^- ions does not change and change is not observed in pH.

$$NaOH \rightleftharpoons Na^+ + OH^-$$

$$NH_4^+ + OH^- \rightleftharpoons NH_4OH$$

Similarly, when a few drops of an acid (e.g. HCl) are added, the H^+ ions from the acid combine with excess of NH_4OH to form H_2O and ammonium ion.

$$HCl \rightleftharpoons H^+ + Cl^-$$

$$NH_4OH + H^+ \rightleftharpoons NH_4^+ + H_2O$$

The H^+ ion concentration does not change and pH remains constant.

2.6.3 Significance

Buffers are highly significant. Biological fluids such as blood, urine etc. have a definite pH. Enzymes act at a specific pH. Any deviation from this normal pH adversely affects the metabolic reactions in the body. Buffer systems help in carrying out the biochemical reactions at normal level. pH is maintained by different buffer systems in the body. Buffers provide protection to the cells and tissues against sudden change in pH. Bicarbonate, phosphate, protein, haemoglobin are some major buffer systems of blood. Carbonic acid-bicarbonate is a common buffer system. H_2CO_3 is a weak acid which dissociates into H^+ and HCO^-.

$$H_2CO_3 \rightleftharpoons H^+ + HCO_3^-.$$

Addition of small amount of an acid does not change the pH of blood, as H^+ from the acid combine with HCO_3^- to form H_2CO_3

$$H^+ + HCO_3^- \rightleftharpoons H_2CO_3.$$

Similarly, if a small amount of a base (NaOH) is added, OH^- ions reacts with H^+ ions to form H_2O

$$NaOH \rightleftharpoons Na^+ + OH^-$$

$$OH^- + H^+ \rightleftharpoons H_2O$$

and the pH is maintained at a constant level.

2.7 ISOMERISM

The compounds which have same molecular formula but different physical and chemical properties are called *isomers* and this phenomenon is known as *isomerism*. The term was given by Berzelius. The difference in properties of the two isomers is due to the difference in the arrangement of atoms within their molecules. The two main types of isomerism are:

1. Structural Isomerism, and
2. Stereoisomerism

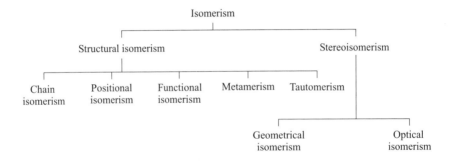

2.7.1 Structural Isomerism

When two compounds have same molecular formula but different structural formula, the phenomenon is called structural isomerism. The structural isomers have same molecular formula but different structures. Structural isomerism may be of the following types:

(a) Chain isomerism

(b) Positional isomerism

(c) Functional isomerism

(d) Metamerism, and

(e) Tautomerism.

(a) *Chain isomerism:* This type of isomerism is due to the difference in the nature of the carbon chain (i.e. straight or branched). For example, butane and isobutane.

$$CH_3 - CH_2 - CH_2 - CH_3 \qquad \overset{\displaystyle CH_3}{\underset{\displaystyle}{CH_3 - CH - CH_3}}$$

n-butane   isobutane

(b) *Positional isomerism:* This type of isomerism is due to the difference in the position of substituent atom or group in the same chain. For example, 1-propanol and 2-propanol are positional isomers.

$$CH_3 - CH_2 - CH_2 - OH \qquad \overset{\displaystyle OH}{\underset{\displaystyle}{CH_3 - CH - CH_3}}$$

1-propanol   2-propanol

(c) *Functional isomerism:* This isomerism is due to difference in nature of functional group present in the isomers. Glyceraldehyde (aldehyde group) and dihydroxy acetone (keto group) are functional isomers.

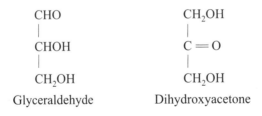

Glyceraldehyde   Dihydroxyacetone

(d) *Metamerism:* In this type of isomerism, the isomers differ in structure due to the difference in distribution of carbon atoms about the functional group. e.g.

$$CH_3 - O - C_3H_7 \qquad\qquad C_2H_5O - C_2H_5$$

Methyl propyl ether   Diethyl ether

(e) *Tautomerism:* It is a special type of isomerism where isomers exist simultaneously in dynamic equilibrium with each other. These isomers are known as tautomers. For example, acetylacetone exists in two forms-keto and enol form.

$$CH_3 - \overset{\overset{O}{\|}}{C} - CH_2 - \overset{\overset{O}{\|}}{C} - CH_3 \rightleftharpoons CH_3 - \overset{\overset{OH}{|}}{C} = CH - \overset{\overset{O}{\|}}{C} - CH_3$$

Keto form   Enol form

2.7.2   Stereoisomerism

When the isomers have the same structural formula but differ in relative arrangement of atoms or groups in space within the molecule, these are known as stereoisomers and the phenomenon is known as stereoisomerism. The spatial arrangement of atoms or groups is also referred as configuration of the molecule. Thus, stereoisomers have the same structural formula but different configuration. Stereoisomerism is of two types:

(a) Geometrical isomerism

(b) Optical isomerism

(a) *Geometrical isomerism:* In geometrical isomerism, the isomers with the same structural formula differ in spatial arrangement of the groups around the double bond. When the similar groups lie on the same side, it is called *cis-isomer* and when similar groups lie on the opposite sides, the compound is *trans-isomer*. For example, maleic acid (cis) and fumaric acid (trans) are geometrical isomers.

Maleic acid (cis)   *Fumaric acid* (trans)

(b) *Optical isomerism:* It is an isomerism where different compounds with same molecular formula, exhibit different optical activity (rotation of plane polarized light). The optical isomerism arises from different arrangement of atoms or groups in three dimensional space resulting in two isomers which are mirror image of each other.

Optical isomers contain an asymmetric (chiral) carbon atom – a carbon atom attached to four different atoms or groups. Because of this asymmetrical carbon atom these isomers show optical activity. e.g.

COOH	COOH	COOH	COOH
H — C — CH$_3$	H$_3$C — C — H	HO — C — H	H — C — OH
NH$_2$	NH$_2$	CH$_2$COOH	CH$_2$COOH
d-alanine	*l-alanine*	*l-malic acid*	*d-malic acid*

Optical isomers have similar chemical and physical properties but differ only in their behaviour towards plane polarized light. The isomer which rotates the plane polarized light towards left is called as *levorotatory* (*l*) and is represented by – sign and the isomer which rotates the plane polarized light to right is called *dextrorotatory* (d) and is represented by +sign. The d- and *l*-forms of a compound are non-superimposable mirror images of each other and are called as *enantiomorphs* or *enantiomores*.

2.8 SUMMARY

- The main theories of acids and bases are—
  Arrhenius Concept, Bronsted-Lowry Concept, and Lewis Concept.
- Diffusion is the movement of molecules from higher concentration to lower concentration down a concentration gradient. Diffusion is a passive movement.
- Osmosis is the movement of solvent from higher concentration to lower concentration across a semipermeable membrane.
- Osmosis is very significant in biological system.
- Viscosity is the measure of fluids' resistance to flow. Viscosity is affected by various factors such as temperature, density and size of the molecules.
- pH is the negative logarithm of hydrogen ion concentration. There are several methods to measure the pH of a solution.
- The solutions which can resist the change in their pH by addition of a small amount of an acid or base, are called buffer solutions. The buffer may be an acid buffer or basic buffer.
- Isomerism is the phenomenon in which two compounds have same molecular formula but different physical and chemical properties. The two main types of isomerism are – structural isomerism and stereoisomerism.

EXERCISE

1. Describe the Bronsted-Lowry concept of acids and bases.
2. Define pH. Explain its significance.
3. Describe the significance of buffer in biological system.
4. Differentiate between:
   (a) Osmosis and diffusion
   (b) Structural and stereoisomerism
5. Write short notes on:
   (a) Viscosity
   (b) Optical isomerism

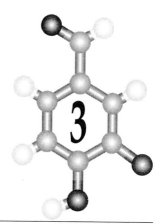

Carbohydrates

Carbohydrates are the most abundant biological molecules. They are aldehyde or ketone compounds with multiple hydroxyl groups. Carbohydrates are the compounds of carbon, hydrogen and oxygen and were originally represented as hydrates of carbon with empirical formula $C_n(H_2O)_n$. However, not all the carbohydrates have this empirical formula.

Carbohydrates are widely distributed in plants and animals. Plants synthesize glucose during photosynthesis from carbon dioxide and water. It is stored in the form of starch or is converted into cellulose which forms the plants' framework.

3.1 BIOLOGICAL SIGNIFICANCE

Carbohydrates have a wide variety of roles in the organisms. They are the major source of energy in the body. Glucose is the most important carbohydrate. The bulk dietary carbohydrates are absorbed in the form of glucose. Carbohydrates (e.g.cellucose) are also the structural elements of plant cell wall. Deoxyribose and ribose sugars form part of the structural frame work of DNA and RNA. Carbohydrates also play role as mediators of cellular interactions.

3.2 CLASSIFICATION

Carbohydrates are classified into three major categories:
1. Monosaccharides
2. Disaccharides
3. Polysaccharides

3.2.1 Monosaccharides

Monosaccharides are the simplest carbohydrates. They are aldehyde or ketone derivatives of straight chain polyhydroxy alcohols containing at least three carbon atoms. On the basis of their carbonyl group, monosaccharides may be either *aldose* (carbonyl group is aldehyde) or *ketose* (carbonyl group is ketone). Monosaccharides cannot be further hydrolyzed into simpler sugars. Depending on the number of carbon atoms, monosaccharides may be *trioses* (n = 3); *tetroses* (n = 4); *pentoses* (n = 5); *hexoses* (n = 6) etc. Dihydroxyacetone and glyceraldehyde are the simplest monosaccharides with 3 carbon atoms. Here, dihydroxyacetone is a ketose as it contains a keto group while, glyceraldehyde is aldose because it contains an aldehyde group.

$$
\begin{array}{cc}
CH_2OH & CHO \\
| & | \\
C=O & HCOH \\
| & | \\
CH_2OH & CH_2OH \\
\text{(Dihydroxyacetone)} & \text{(Glyceraldehyde)} \\
\text{(a ketose)} & \text{(an aldose)}
\end{array}
$$

Glucose is a hexosugar with six carbon atoms ($C_6H_{12}O_6$). It is an aldose as it contains an aldehyde group, the keto-hexose sugar is fructose.

$$
\begin{array}{cc}
CHO & CH_2OH \\
| & | \\
H-C-OH & C=O \\
| & | \\
HO-C-H & HO-C-H \\
| & | \\
H-C-OH & H-C-OH \\
| & | \\
H-C-OH & H-C-OH \\
| & | \\
CH_2OH & CH_2OH \\
\textit{Glucose} & \textit{Fructose}
\end{array}
$$

The aldehyde and ketone groups of carbohydrates with five and six carbon atoms react with alcohol group present in the neighbouring carbons to form *hemiacetal* and *hemiketals* respectively. This results in the formation of five or six membered ring.

$$
\underset{\textit{Alcohol}}{R-OH} + \underset{\textit{Aldehyde}}{R'-C\overset{H}{\underset{\diagdown O}{\diagup}}} \rightleftharpoons \underset{\textit{Hemiacetal}}{\overset{OR\quad H}{\underset{R'\quad OH}{C}}}
$$

$$
\underset{\textit{Alcohol}}{R-OH} + \underset{\textit{Ketone}}{R'-C\overset{R''}{\underset{\diagdown O}{\diagup}}} \rightleftharpoons \underset{\textit{Hemiketal}}{\overset{OR\quad R''}{\underset{R'\quad OH}{C}}}
$$

The five membered ring structure resembles the organic molecule *furan* and thus the derivatives with this structure are termed *furanoses*. Those with six-membered ring resemble the organic molecule *pyran* and thus they are termed *pyranoses*.

Pyran   *Furan*

The cyclic form of glucose with six-membered ring is known as *glucopyranose* and that of fructose with five membered ring is called *fructofuranose*.

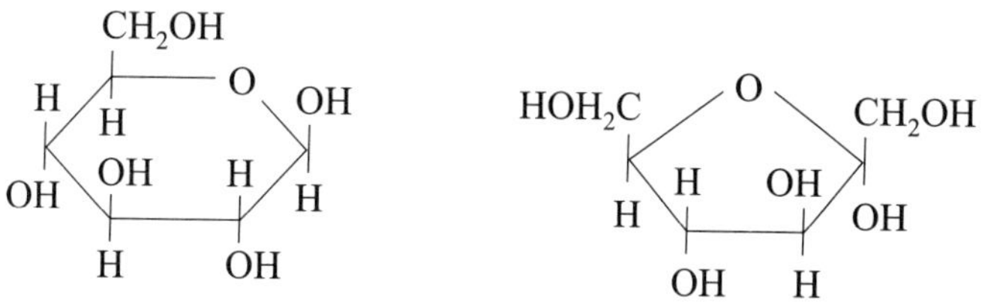

β-D-*Glucopyranose*   α-D-*Fructofuranose*

Table 3.1   Some monosaccharides

Carbohydrates	No. of Carbon Atoms	Aldoses	Ketoses
Trioses ($C_3H_6O_3$)	3	Glycerose	Dihydroxyacetone
Tetroses ($C_4H_8O_4$)	4	Erythrose	Erythrulose
Pentoses ($C_5H_{10}O_5$)	5	Ribose	Ribulose
Hexoses ($C_6H_{12}O_6$)	6	Glucose	Fructose
Heptoses ($C_7H_{14}O_7$)	7	Mannoheptulose	Sedoheptulose

3.2.2   Disaccharides

Disaccharides are those carbohydrates which yield two molecules of monosaccharides on hydrolysis e.g. Maltose (glucose and glucose); sucrose (glucose and fructose); lactose (glucose and galactose).

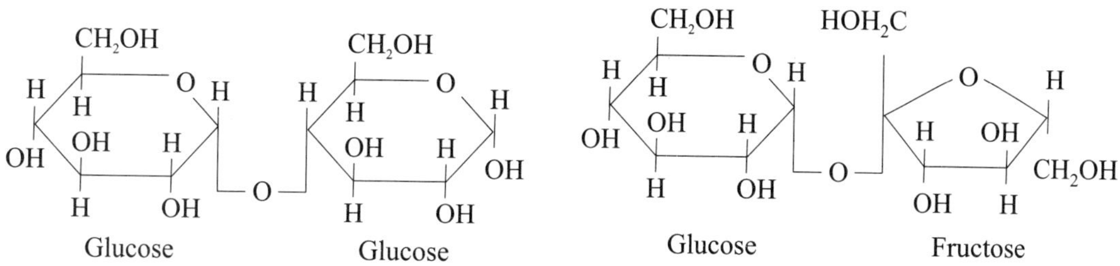

Glucose   Glucose

Maltose

Glucose   Fructose

Sucrose

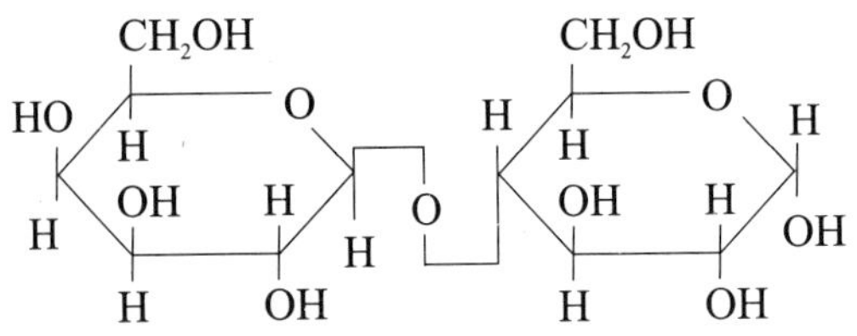

Galactose   Glucose

Lactose

Sucrose is the most abundant disaccharide. It is obtained commercially from cane or beet. It can be cleaved into its components – glucose and fructose by the enzyme *sucrase*. Lactose occurs naturally only in milk. It consists of galactose joined to glucose by a β-1,4 glucosidic linkage. It is hydrolyzed to its monosaccharides by enzyme *lactase* in human beings. In maltose, two glucose units are joined by a α-1,4 glycosidic linkage. Enzyme *maltase* can hydrolyze it to glucose.

3.2.3 Polysaccharides

Polysaccharides yield more than two molecules of monosaccharides on hydrolysis. The most frequent monosaccharide units in polysaccharides is glucose, although fructose, galactose and other hexoses also occur. If all the monosaccharides are the same in a polysaccharide, it is called a *homopolysaccharide*. For example, starch, cellulose and glycogen are the common homopolysaccharides. If the monosaccharides are different in a polysaccharide, it is called a *heteropolysaccharide*. Hyaluronic acid, pectin and heparin are the examples of hetropolysaccharides.

3.2.3.1 *Cellulose*

Cellulose is the primary structural component of plant cell wall. It is one of the most abundant organic compounds in the biosphere. Cellulose is a linear polymer of upto 15,000 D-glucose residues linked by β-1,4 linkages. The β-configuration allows cellulose to form very long, straight chains. The straight chain formed by β-linkages is optimal for the construction of fibres having a high tensile strength.

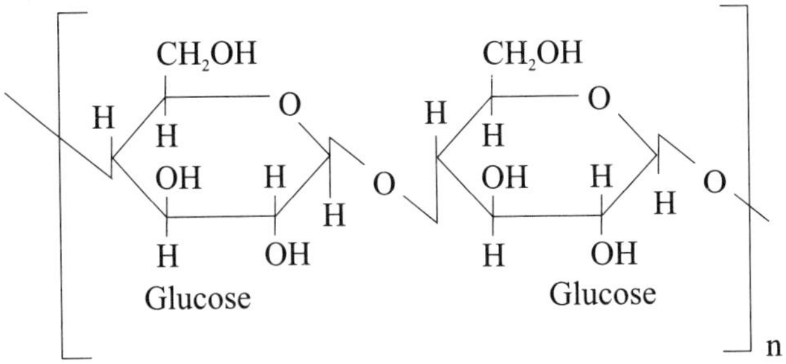

Cellulose

3.2.3.2 *Starch*

Starch is the reserve food material in the plants. It is a homopolysaccharide made up of glucose units. The glucose residues are arranged in the form of branched and unbranched chain. In starch, glucose residues are linked by α-1,4 glycosidic linkages. The branches are formed by α-1,6 glycosidic linkages. Unbranched starch is called amylose and the branched starch is called amylopectin. α-amylose is a linear polymer of several thousand glucose residues linked by α-$(1\rightarrow4)$ bonds. Amylopectin consists mainly of α-$(1\rightarrow4)$ linked glucose residues but is a branched molecule with α $(1\rightarrow6)$ branch points every 24 to 30 glucose on average.

α-*Amylose*

3.2.3.3 *Glycogen*

Glycogen is the major reserve food material in animals. Glycogen is a homopolysaccharide and yields glucose on complete hydrolysis. It is a branched polymer of glucose in which glucose residues are linked by α-1,4 glucosidic and α-1,6 glycosidic linkages. Glycogen is stored mainly in the liver and muscles and is changed into glucose when the blood glucose level falls. The primary structure of glycogen resembles that of amylopectin, but glycogen is more highly branched with branch points occuring every 8 to 12 glucose residues.

Amylopectin

3.2.3.4 *Chitin*

Chitin is the structural component of exoskeleton of invertebrates such as insects and crustaceans. It is also present in the cell walls of fungi and many algae. Chitin is a homopolysaccharide made up of N-acetyl-D-glucosamine linked by β-1,4 glycosidic linkage. Chitin is decomposed to N-acetyl glucosamine by *chitinase* present in the gastric juices of snails or from bacteria.

N-acetyl glucosamine   N-acetyl glucosamine

Chitin

3.3 PROPERTIES OF CARBOHYDRATES

3.3.1 Physical Properties of Monosaccharides

Monosaccharides are colourless and crystalline compounds. They are sweet in taste and are soluble in water, spraingly soluble in alcohol and insoluble in ether. They contain asymmetric carbon atom and hence exist in two forms which are mirror images of each other.

Optical Activity: They are optically active compounds. They rotate the plane polarized light. When a polarized light is passed through a solution of these carbohydrates, the plane of light is rotated either right or left. Carbohydrates isolated from living systems are a mixture of several isomers. The degree of optical rotation may change due to inter conversion of isomers. A freshly prepared aqueous solution of α-D glucose has a specific rotation of +112.2°. When this solution is allowed to stand, the rotation falls to +52.7° and remains constant at this value. This gradual change in specific rotation is called *mutarotation*.

3.3.2 Chemical Properties

The chemical properties are based on the active groups present in the monosaccharides.

3.3.2.1 Oxidation

Monosaccharides are easily oxidized, thus act as best reducing agents. They readily reduce oxidizing agents such as ferricyanides, hydrogen peroxide or cupric ion. Glucose reduces Tollen's reagent, Fehling solution and Benedict's reagent. These reactions are used in the estimation of glucose and other sugars. In these reactions glucose is oxidized to gluconic acid.

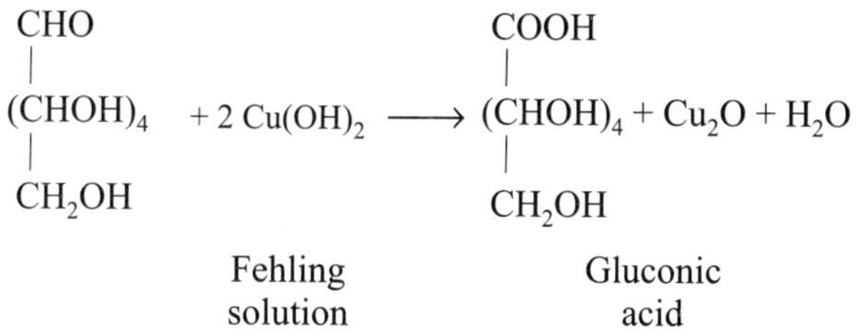

$$\begin{array}{l}\text{CHO}\\|\\(\text{CHOH})_4\\|\\\text{CH}_2\text{OH}\end{array} + 2\,\text{Cu(OH)}_2 \longrightarrow \begin{array}{l}\text{COOH}\\|\\(\text{CHOH})_4\\|\\\text{CH}_2\text{OH}\end{array} + \text{Cu}_2\text{O} + \text{H}_2\text{O}$$

Fehling solution

Gluconic acid

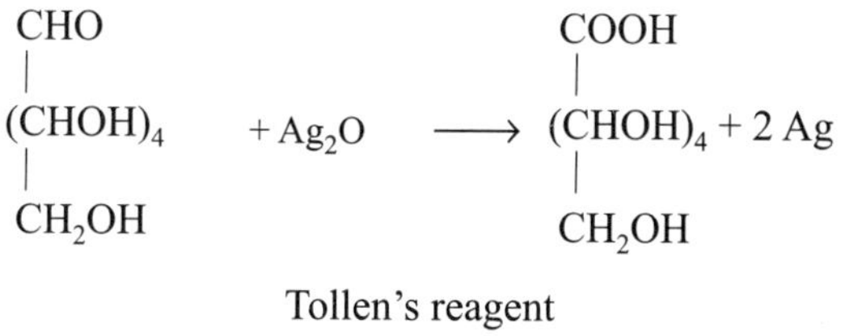

$$\begin{array}{l}\text{CHO}\\|\\(\text{CHOH})_4\\|\\\text{CH}_2\text{OH}\end{array} + \text{Ag}_2\text{O} \longrightarrow \begin{array}{l}\text{COOH}\\|\\(\text{CHOH})_4\\|\\\text{CH}_2\text{OH}\end{array} + 2\,\text{Ag}$$

Tollen's reagent

3.3.2.2 *Reduction*

Monosaccharides can be reduced by various reducing agents. The reduction is due to the presence of —CHO or =CO group. On reduction, they yield alcohols. With sodium glucose yields sorbitol.

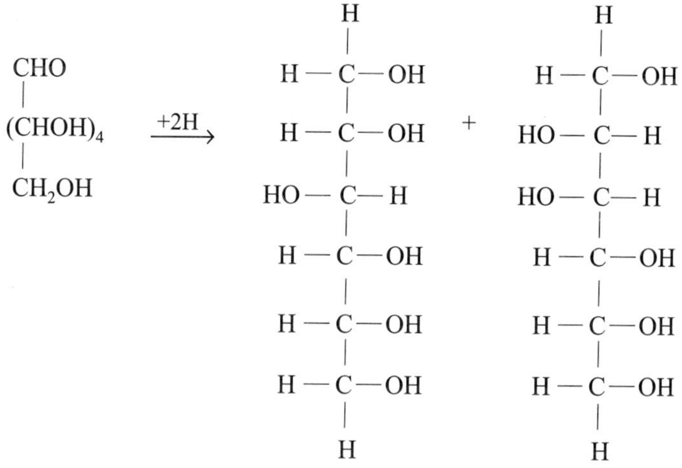

$$\begin{array}{l}\text{CHO}\\|\\(\text{CHOH})_4\\|\\\text{CH}_2\text{OH}\end{array} \xrightarrow{+2\text{H}} \text{Sorbitol} + \text{Mannitol}$$

Glucose   Sorbitol   Mannitol

3.3.2.3 *Esterification*

Glucose reacts with five molecules of acetic anhydride in the presence of pyridine or anhydrous $ZnCl_2$ to from acetyl derivatives (esters). This ability is due to the presence of hydroxyl group.

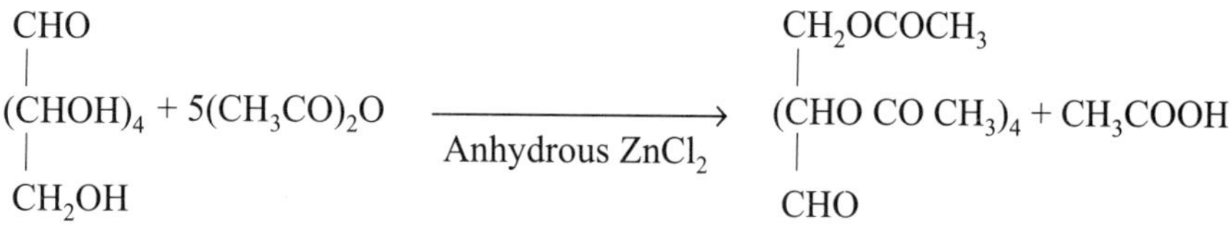

$$\begin{array}{l}\text{CHO}\\|\\(\text{CHOH})_4\\|\\\text{CH}_2\text{OH}\end{array} + 5(\text{CH}_3\text{CO})_2\text{O} \xrightarrow{\text{Anhydrous ZnCl}_2} \begin{array}{l}\text{CH}_2\text{OCOCH}_3\\|\\(\text{CHO CO CH}_3)_4\\|\\\text{CHO}\end{array} + \text{CH}_3\text{COOH}$$

Penta-acetyl glucose

3.3.2.4 Condensation Reaction

Aldyhyde groups of monosaccharides condense with primary amines to form a Schiff's base.

$$
\begin{array}{ccccc}
\text{CHO} & & \text{CH}_3 & & \text{H} \quad\quad \text{CH}_3 \\
| & & | & & | \quad\quad\quad | \\
(\text{CHOH})_4 & + & \text{CH(NH}_2) & \longrightarrow & \text{C}\!=\!\text{N}\!-\!\text{CH} \; + \; \text{H}_2\text{O} \\
| & & | & & | \quad\quad\quad | \\
\text{CH}_2\text{OH} & & \text{COOH} & & (\text{CHOH})_4 \; \text{COOH} \\
& & & & | \\
& & & & \text{CH}_2\text{OH}
\end{array}
$$

Alanine   Schiff's base

3.3.2.5 Fermentation

Fermentation involves the conversion of a large molecule into simple molecules by means of enzymes in an anaerobic condition. It releases energy as well as CO_2. Glucose gives ethyl alcohol and CO_2 during fermentation by yeast.

$$C_6H_{12}O_6 \longrightarrow 2C_2H_5OH + CO_2$$

Glucose   Ethyl alcohol

3.4 SUMMARY

- Carbohydrates are the compounds of carbon, hydrogen and oxygen and are the major source of energy in body.
- Carbohydrates are classified into three major groups – monosaccharides, disaccharides and polysaccharides.
- Monosaccharides are the simplest carbohydrates and are either aldoses or ketoses.
- Disaccharides consist of two molecules of monosaccharide.
- In polysaccharides a number of monosaccharides are linked by glycosidic bonds.
- Polysaccharides may be either homopolysaccharides or heteropolysaccharides.
- Monosaccharides are colourless and crystalline compounds which are optically active.
- The chemical properties of monosaccharides depend on the active groups (—CHO or =CO) present in them.

EXERCISE

1. What are carbohydrates? Discuss the physical and chemical properties of carbohydrates.
2. Give an account of biological significance of carbohydrates.
3. Write the structure and significance of the following:
   (a) Cellulose
   (b) Starch

Lipids

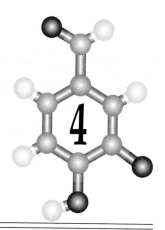

Lipids are a very important heterogenous group of organic substances which are widely distributed throughout the plant and animal kingdom. In plants, they are present in seeds, nuts and fruits, while in the animals they are stored in adipose tissues, bone marrow and nervous tissues. Chemically they are various types of esters of fatty acids with alcohols. In addition to alcohols and fatty acids, some of the lipids may contain phosphoric acid, nitrogenous group and carbohydrates. *Bloor* defined lipids as "naturally occurring compounds which are soluble in one or more organic solvents such as benzene, chloroform, ether and acetone and so called fat solvents and on hydrolysis yield fatty acids which are utilized by living organisms."

4.1 BIOLOGICAL SIGNIFICANCE

Lipids are important dietary constituents and serve as a source of energy. Lipids yield more energy per gm (9.5 cal/gm) as compared to carbohydrates (4 cal/gm). Lipids are also the building material in cellular components. Lipoproteins and phospholipids are important constituents of biological membranes. The nervous system is particularly rich in lipids and they are essential for the proper functioning of the nervous system.

In animals, fat is deposited as subcutaneous fat and provides an insulating effect in the body. Lipids, around many vital organs, protect them against mechanical injury.

Waxes also form protective covering in plants and animals. They protect living surface against bacteria and insect invasion.

4.2 CLASSIFICATION

Lipids are generally classified into the following groups:
1. Simple Lipids
2. Compound Lipids, and
3. Derived Lipids

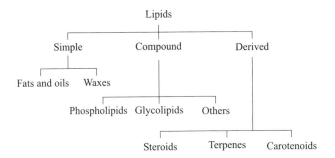

4.2.1 Simple Lipids

These are the esters of fatty acids with various alcohols. These can be further categorized into *fats* and *waxes*.

Fats are the esters of fatty acids with glycerol. A fat in the liquid state is called *oil*. Waxes are the esters of fatty acids with higher molecular weight monohydric alcohols.

4.2.1.1 Fats

Fats are solids at room temperature. Chemically fats are triglycerides since three molecules of fatty acids condense with one molecule of glycerol e.g. Three molecules of butyric acid are linked to glycerol to yield tributyrin – a fat.

$$
\begin{array}{lll}
CH_2OH & COOH\ CH_2{\cdot}CH_2\ CH_3 & CH_2O{\cdot}CO{\cdot}CH_2\ CH_2\ CH_3 \\
| & & | \\
CHOH & +\ COOH{\cdot}CH_2{\cdot}CH_2CH_3 \longrightarrow & CHO{\cdot}CO{\cdot}CH_2CH_2CH_3 + 3H_2O \\
| & & | \\
CH_2OH & COOH{\cdot}CH_2CH_2CH_3 & CH_2OCOCH_2CH_2\ CH_3
\end{array}
$$

Glycerol   Butyric acid   Tributyrin

If all the three molecules of fatty acids are similar, it is a *simple glyceride*. If fatty acid molecules are different it is called a *mixed glyceride*. Natural fats are largely composed of mixed glycerides.`

The melting point of fat depends upon the chain length and degree of saturation of fatty acids. The melting points of fats are always higher than their solidification points. Fats develop unpleasant odour on ageing. This is due to autooxidation of fat. This is called *rancidification*. The chemical changes that occur during rancidification are called *rancidity*.

Fats which are liquid at room temperature are called *oils*. Oils are also esters of fatty acids and glycerol. But the fatty acids found in oils are unsaturated fatty acids. The unsaturated fatty acids have one or more double bonds. They have low melting point and are insoluble in water.

4.2.1.2 Waxes

Waxes are the esters of fatty acids with high molecular weight alcohol. Being highly insoluble in water and having no double bonds in their hydrocarbon chains, waxes are chemically inert. They are resistant to atmospheric oxidation. Waxes serve as protective coatings on fruits and leaves. They serve admirably on leaf surfaces to protect plants from water loss and abrasive damage. Waxes play an important role in providing a water barrier for insects, birds and animals such as sheep. Some common waxes are *bees wax, carnauba wax, sperm whale wax* and *ambretolide*. The bees-wax contains palmitic acid and myricyl alcohol and is called *myricyl palmitate*. The sperm whale wax contains palmitic acid and cetyl alcohol and is called *cetyl palmitate*. Carnauba wax contains fatty acids esterified with tetracosanol and tetratriacontanol. Ambretolide found in the seeds of *Abelmoschus esculentus* is a hydroxyl acid and is responsible for the charactristic smell of the seed.

4.2.1.3 Fatty Acids

Fatty acids are obtained from the hydrolysis of fats. A fatty acid can be defined as an organic acid that occurs in a natural triglyceride and is a monocarboxylic acid ranging in chain length from C_4 to about 24 carbon atoms. Fatty acids that occur in natural fats usually contain an even number of carbon atoms and are straight chain derivatives. The fatty acids can be divided into different categories on different basis:

On the basis of *presence or absence of double bonds*, fatty acids may be—

(a) Saturated, and

(b) Unsaturated fatty acids

(a) *Saturated fatty acids:* The fatty acids which *do not contain* any double bond are called saturated fatty acids. (Table 4.1)

The general formula is $C_nH_{2n+1}COOH$ e.g. Butyric acid $CH_3(CH_2)_2COOH$.

Table 4.1 Some naturally occurring saturated fatty acids (Straight chain acids)

Carbon Atoms	Common Name	Systematic Name	Structure	Common Sources
4	Butyric acid	Butanoic acid	$CH_3(CH_2)_2COOH$	Butter
6	Caproic acid	Hexanoic acid	$CH_3(CH_2)_4COOH$	Coconut and palm oil
12	Lauric acid	Dodecanoic acid	$CH_3(CH_2)_{10}COOH$	Laurel oil, spermaceti
14	Myristic acid	Tetradecanoic acid	$CH_3(CH_2)_{12}COOH$	Butter and wool fats
16	Palmitic acid	Hexadecanoic acid	$CH_3(CH_2)_{14}COOH$	Animal and plant fats
18	Stearic acid	Octadecanoic acid	$CH_3(CH_2)_{16}COOH$	Animal and plant fats
20	Arachidic acid	Eicosanoic acid	$CH_3(CH_2)_{18}COOH$	Ground nut oil
24	Lignceric acid	Tetracosanoic acid	$CH_3(CH_2)_{22}COOH$	Ground nut and rapeseed oil
26	Cerotic acid	Hexacosanoic acid	$CH_3(CH_2)_{24}COOH$	Wool fat

The saturated fatty acids are straight chain acids. In addition to these straight chain acids, there are some branched chain acids, with odd or even number of carbon atoms e.g. Isopalmitic acid, ante-isopalmitic acid, tuberculostearic acid (Table 4.2).

Table 4.2 Some naturally occurring saturated (branched chain) fatty acids

Carbon Atoms	Common Name	Systematic Name	Structure	Common Source
16	Isopalmitic acid	Isohexadecanoic acid	CH_3 $CH(CH_2)_{12}COOH$ CH_3	Wool fat
17	Ante-isopalmitic acid	14-methyl hexadecanoic acid	CH_3CH_2 $CH(CH_2)_{12}COOH$ CH_3	Wool fat
19	Tuberculo-stearic acid	D(-)10-methyl octadecanoic acid	CH_3 | $CH_3(CH_2)_7CH(CH_2)_8COOH$	Bacteria

(b) *Unsaturated fatty acids:* The unsaturated fatty acids have one or more double bonds. On the basis of number of double bonds the unsaturated fatty acids may be either *monounsaturated* (one double bond) e.g. crotonic acid, oleic acid, palmitoleic acid, nervonic acid etc. or *polyunsaturated fatty acids* (more than one double bond) e.g. linoleic acids, eleostearic acid etc.

In most of the monounsaturated fatty acids there is a single double bond lying between carbon atom 9 and 10. This is designated as Δ^9. The symbol Δ with the superscript number 9 indicates the position of double bond. When there are more than one double bond (Polyunsaturated fatty acids), the additional bonds occur between the Δ^9 double bond and the methyl terminal end of the chain. The symbol 18:3 signifies that there are three double bonds. The symbol 18:0 denotes a C_{18} fatty acid with no double bond. (Table 4.3).

Geometrical isomerism in unsaturated fatty acids: Geometrical isomerism is present in unsaturated fatty acids due to the presence of C=C. Two isomers are possible depending upon the spatial arrangements of groups or atoms around the C=C. They are *cis* and *trans* isomers. In cis isomers identical groups are present on the same side of the C=C. In trans isomers identical groups are occupying opposite sides of C=C e.g. oleic acid and elaidic acid.

Oleic acid
(cis)

Elaidic acid
(trans)

Table 4.3 Some naturally occurring unsaturated fatty acids

Carbon Skeleton	Common Name	Systematic Name	Structure	Common Source
$16:1\ (\Delta^9)$	Palmitoleic acid	9-hexadecenoic acid	$CH_3(CH_2)_5CH = CH(CH_2)_7COOH$	Animal and plant fats
$18:1\ (\Delta^9)$	Oleic acid	9-octadecenoic acid	$CH_3(CH_2)_7CH = CH(CH_2)_7COOH$	Animal and plant fats
$18:2\ (\Delta^{9,12})$	Linoleic acid	9,12-octadecadienoic acid	$CH_3(CH_2)_4CH = CHCH_2CH = CH(CH_2)_7COOH$	Linseed and cotton seed
$18:3\ (\Delta^{9,12,15})$	α-Linoleic acid	9,12,15-octadecatrienoic acid	$CH_3CH_2CH = CHCH_2CH = CHCH_2CH = CH(CH_2)_7COOH$	Linseed oil
$20:4\ (\Delta^{5,8,11,14})$	Arachidonic acid	5,8,11,14-icosatetraenoic acid	$CH_3(CH_2)_4CH = CHCH_2CH = CHCH_2CH = CHCH_2CH = CH(CH_2)_3COOH$	Animal fat

Hydroxy or oxygenated fatty acids: Fatty acids containing hydroxyl groups are called hydroxy fatty acids e.g. cerebronic acid of brain glycolipids, ricinoleic acid in castor oil.

$$CH_3 - (CH_2)_{21} - \underset{\underset{OH}{|}}{CH} - COOH \qquad \textit{Cerebronic acid}$$

$$CH_3 - (CH_2)_5 - \underset{\underset{OH}{|}}{CH} - CH_2 - CH = CH(CH_2)_7COOH \qquad \textit{Ricinoleic acid}$$

Cyclic fatty acids: Fatty acids bearing cyclic groups e.g. chaulmoogric acid, hydnocarpic acid, lactobacillic acid, sterculic acid.

$$\begin{array}{c} CH = CH \\ | \qquad \ \ \diagdown \\ CH_2 - CH_2 \diagup \end{array} CH(CH_2)_{12}COOH \qquad \textit{Chaulmoogric acid}$$

$$\begin{array}{c} CH = CH \\ | \qquad \ \ \diagdown \\ CH_2 - CH_2 \diagup \end{array} CH(CH_2)_{10}COOH \qquad \textit{Hydnocarpic acid}$$

$$CH_3 - (CH_2)_5 - \overset{\overset{\displaystyle CH_2}{\diagup \ \diagdown}}{CH - CH} - (CH_2)_9\ COOH \qquad \textit{Lactobacillic acid}$$

$$CH_3(CH_2)_7 - \overset{\overset{\displaystyle CH_2}{\diagup \ \diagdown}}{C = C} - (CH_2)_7 - COOH \qquad \textit{Sterculic acid}$$

4.2.1.4 *Essential and Non-Essential Fatty Acids*

Essential fatty acids: The fatty acids which cannot be synthesized by human body but are essential for the normal maintenance of the body are called essential fatty acids. These fatty acids must be included in our diet. Three polyunsaturated fatty acids, linoleic acid, *linolenic acid* and *arachidonic acid* are the essential fatty acids. Essential fatty acids play various important roles in the body and lack of these in the diet can produce growth retardation and other deficiency manifestation symptoms in animals.

Non-essential fatty acids: These are the fatty acids which can be synthesized by our body. Thus they need not be included in our diet. They are unsaturated fatty acids and are synthesized from their corresponding saturated fatty acids by introducing a single double bond e.g. polmitoleic acid and oleic acid.

4.2.2 Compound Lipids

Compound lipids are the esters of fatty acids containing groups other than and in addition to an alcohol and fatty acids.

Compound lipids can be categorized into the following types:

- Phospholipids,
- Glycolipids, and
- Other Compound lipids

4.2.2.1 *Phospholipids*

Phospholipids are those compound lipids which contain a phosphorus atom. Phospholipids are wide spread in bacteria, animal and plant tissues and their general structures are quite similar. These have been termed as *amphipathic* compounds since they possess both polar and non-polar functions.

$$CH_2OCO\ R_1$$
$$|$$
$$CHOCO\ R_2$$
$$|\qquad\qquad OH$$
$$|\qquad\qquad |$$
$$CH_2 - O - P = O$$
$$|$$
$$OH$$

(Phospholipid)

In addition to phosphorus, phospholipids may also contain nitrogen as a key component. There are various types of phospholipids:

Lecithin: Lecithin is widely distributed in nature. In animals it is found in liver, brain, nerve tissues, sperm and egg-yolk. In plants it is abundant in seeds and sprouts. On hydrolysis, lecithin yields glycerol, fatty acids, phosphoric acid and nitrogenous base choline. Lecithin is also called *phosphatidyl choline*. The fatty acids commonly found in lecithin are palmitic, stearic, oleic, linolenic and arachidonic acids. Lecithin is a yellowish grey solid which is soluble in ether and alcohol.

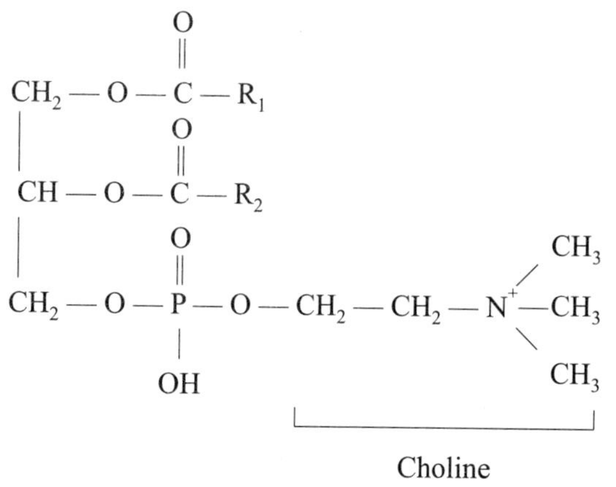

$$CH_2 - O - \overset{\displaystyle O}{\overset{\displaystyle \|}{C}} - R_1$$

$$CH - O - \overset{\displaystyle O}{\overset{\displaystyle \|}{C}} - R_2$$

$$CH_2 - O - \overset{\displaystyle O}{\overset{\displaystyle \|}{\underset{\displaystyle OH}{P}}} - O - CH_2 - CH_2 - \overset{+}{N} \begin{matrix} CH_3 \\ - CH_3 \\ CH_3 \end{matrix}$$

Choline

Lecithin (Phosphatidyl choline)

Cephalins: Cephalins are found in animal tissues in close association with lecithins. The basic difference between cephalins and lecithins is the nature of nitrogenous base. Cephalins contain ethanolamine in place of choline. Some times serine is present in place of choline. The fatty acid components of cephalins are stearic, oleic, linoleic and arachidonic acid. They are less soluble in alcohol than lecithins.

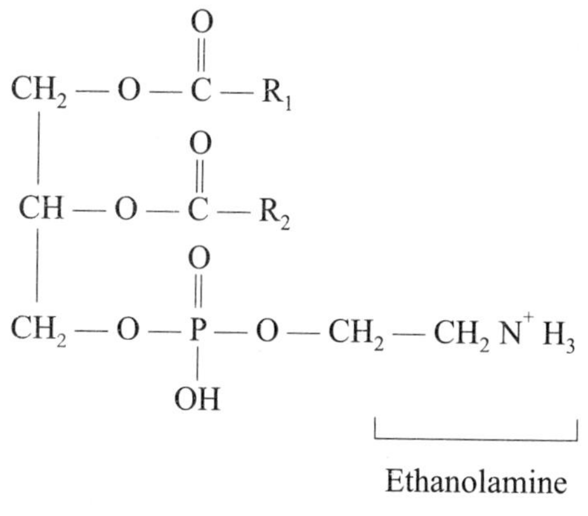

$$CH_2 - O - \overset{\displaystyle O}{\overset{\displaystyle \|}{C}} - R_1$$

$$CH - O - \overset{\displaystyle O}{\overset{\displaystyle \|}{C}} - R_2$$

$$CH_2 - O - \overset{\displaystyle O}{\overset{\displaystyle \|}{\underset{\displaystyle OH}{P}}} - O - CH_2 - CH_2 \overset{+}{N} H_3$$

Ethanolamine

Cephalin (Phosphatidyl ethanolamine)

Plasmalogens: Plasmalogens are abundant in brain and muscles. They are also present in seeds of higher plants. Structurally, they resemble lecithins and cephalins except in having an aldehyde group attached to α-carbon atom of glycerol. They are soluble in all lipid solvents.

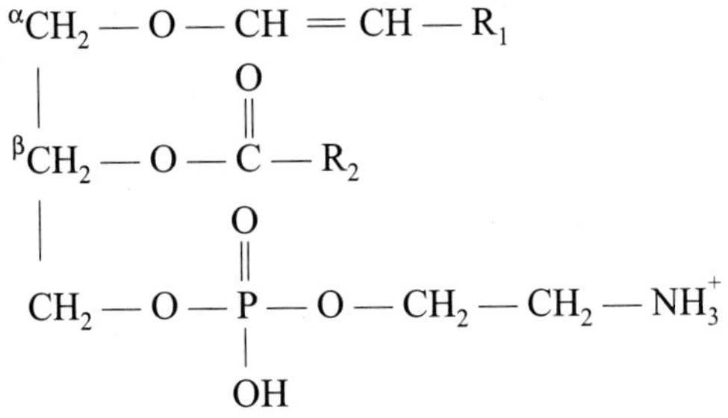

$$^{\alpha}CH_2 - O - CH = CH - R_1$$

$$^{\beta}CH_2 - O - \overset{\displaystyle O}{\overset{\|}{C}} - R_2$$

$$CH_2 - O - \overset{\displaystyle O}{\underset{\displaystyle OH}{\overset{\|}{P}}} - O - CH_2 - CH_2 - NH_3^+$$

Plasmalogens

Phosphoinositides: They are present in brain tissues and nervous tisssues. These phospholipids contain hexahydric alcohol inositol

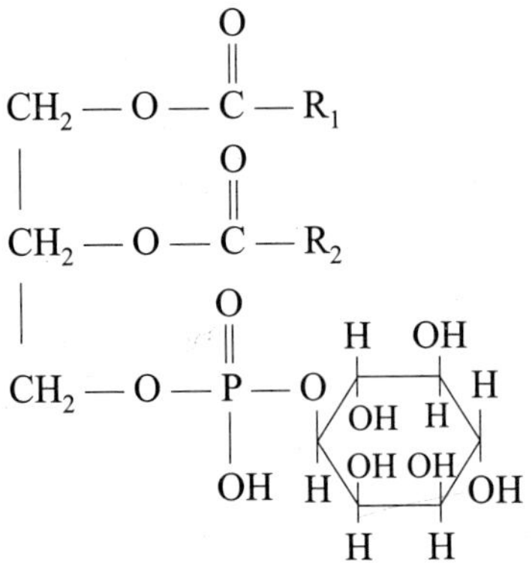

$$CH_2 - O - \overset{\displaystyle O}{\overset{\|}{C}} - R_1$$

$$CH_2 - O - \overset{\displaystyle O}{\overset{\|}{C}} - R_2$$

$$CH_2 - O - \overset{\displaystyle O}{\underset{\displaystyle OH}{\overset{\|}{P}}} - O$$

Phosphoinositide (Phosphatidyl inositol)

Phosphosphingosides: These are commonly found in nerve tissues especially in the myelin sheath of the nerves. In these lipids glycerol is replaced by an 18 carbon unsaturated amino alcohol called 'sphingosine.' On hydolysis phosphosphingosides yield fatty acid, phosphoric acid, choline and sphingosine.

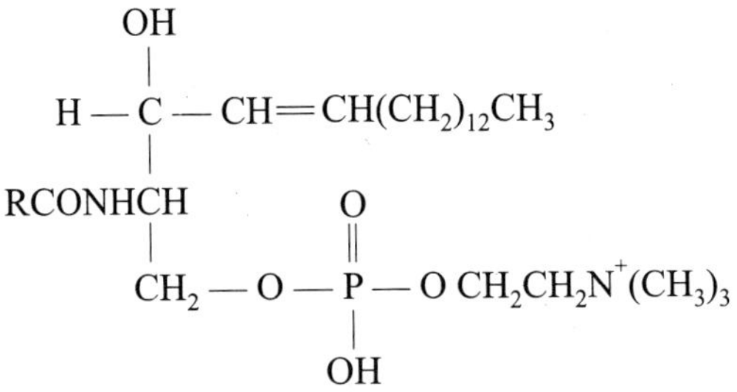

$$\overset{\displaystyle OH}{\underset{\displaystyle RCONHCH}{H - \overset{|}{\underset{|}{C}} - CH = CH(CH_2)_{12}CH_3}}$$

$$CH_2 - O - \overset{\displaystyle O}{\underset{\displaystyle OH}{\overset{\|}{P}}} - O \ CH_2CH_2N^+(CH_3)_3$$

Sphingomyelin

4.2.2.2 Glycolipids

Glycolipids are compounds containing carbohydrates and high molecular weight fatty acids like sphingosine but no phosphoric acid. These are of two types:

   (i) Cerebrosides

  (ii) Gangliosides

Cerebrosides: Cerebrosides occur in large amount in brain and myelin sheath of nerves. The structure of cerebrosides is some what similar to sphingomyelin. Here the fatty acid-ceramide is linked either to galactose or glucose.

Cerebroside

Gangliosides: Gangliosides are found in ganglion cells of nervous tissues and also in parenchymatous tissues like spleen and RBCs. They are the most complex glycosphingolipids. They are ceramides with attached oligosaccharides that include at least one sialic acid residue. They are primarily components of cell surface membrane. Gangliosides have considerable physiological and medical significance. They are the receptors for certain bacterial protein toxins such as cholera toxins and certain pituitary glycoprotein hormones.

4.2.2.3 Other Compound Lipids

Besides, the above mentioned lipids, there are lipids containing sulfur, known as Sulfolipids. Sulfolipids are widely distributed in plants and are localized in chloroplast. They are also found in the chromatophores of photosynthetic bacteria.

    Lipoprotein is another group of compound lipids. Lipoproteins are the components of membranes. They are found in the membranes of mitochondria, endoplasmic reticulum and nuclei. The electron transport chain system in mitochondria appears to contain large amounts of lipoproteins. The lipid components consist of triacylglycerol, phospholipid and cholesterol. The protein components of lipoprotein have a relatively high portion of non-polar amino acid residues that can participate in the binding of the lipids.

4.2.3 Derived Lipids

Derived lipids are the products of hydrolysis of simple lipids and compound lipids and include compounds like steroids, terpenes, fatty acids, alcohols, aldehydes and carotenoids.

4.2.3.1 Steroids

Steroids are the derivatives of *cyclopentanoperhydrophenanthrene*, a compound consisting of four fused non-planer rings. They are named as A, B, C and D. The rings A, B and C are hexagons and are called *cyclohexane rings* while D is a pentagon and is called *cyclopentane*. The 3 cyclohexane rings are fused in a non-linear or phenanthrene manner and the cyclopentane ring is fused terminally.

Cyclopentanoperhydrophenanthrene

The steroids may have one or more alcoholic groups. The steroids having alcoholic groups are called *sterols*. Sterols are crystalline compounds and differ from common alcohols in being solids and that is the reason they are called sterols. *Cholesterol*, a major component of animal plasma membrane is classified as sterol because of its C3OH group. Cholesterol is a crystalline solid with high melting point.

Cholesterol

In mammals, cholesterol is the metabolic precursor of steroid hormones— adrenocorticoids and sex-steroids. Adrenocorticoids which include glucocorticoids, aldosterone and other mineralocorticoids influence a wide variety of vital functions. Sex-steroids (androgens, estrogens and progeotrone) affect sexual development and functions.

4.2.3.2 Terpenes

Terpenes are a group of derived lipids which are found in plants. Chemically they are hydrocarbons which are constructed out of *isoprene* units. Each terpene has 5 carbon atoms and 8 hydrogen atoms. Each *terpene has* two or more *isoprene* units. Based on number of isoprene units, terpenes are classified into six types:

1. *Monoterpenes* contain two units of isoprene e.g. *menthol, limonene*. The monoterpenes are used as perfumes.
2. *Sequiterpenes* contain three units of isoprene and are also used as perfumes e.g. *farnesol*.
3. *Diterpenes* contain four units of isoprene and are found as substituents of the resins and balsams e.g. abietic acid.

4. *Triterpenes* contain six isoprene units. They are produced as intermediate products during the biosynthesis of cholesterol e.g. lanosterol, a major constituent of wool and fat.

5. *Tetraterpenes* contain 8 isoprene units e.g. carotenoids.

6. *Polyterpenes* contain more than eight units of isoprene units e.g. rubber.

$$CH_2 = Cl - \overset{\overset{\displaystyle CH_3}{|}}{C}H = CH_2$$

Isoprene (basic unit of terpenes)

4.2.3.3 Carotenoids

Carotenoids are a group of derived lipids which are exclusively of plant origin. They are isoprene derivatives. They are tetraterpenes containing eight isoprene units e.g. lycopene, carotene, xanthophylls.

4.3 SUMMARY

- Lipids are a heterogenous group of molecules which are soluble in organic solvents.
- Chemically, lipids are the esters of fatty acids with alcohols.
- Lipids are good sources of energy and important constituents of biological membranes.
- Lipids are generally classified into simple, compound and derived lipids.
- Fats and waxes are the simple lipids.
- Fatty acids are obtained from the hydrolysis of fats and can be saturated or unsaturated.
- Phospholipids are the compound lipids which contain a phosphorus group.
- Glycolipids contain carbohydrates and high molecular weight fatty acids.
- Derived lipids are the products of hydrolysis of simple and compound lipids and include steroids, terpenes, carotenoids etc.

EXERCISE

1. Explain – 'Lipids are a heterogenous group of chemical substances'.
2. Write the biological significance of lipids.
3. Differentiate between:
   (a) Fat and oil
   (b) Saturated and unsaturated fatty acids.
4. How does unsaturation affect the physical properties of fatty acids?
5. Write notes on:
   (a) Cephalins
   (b) Cerebrosides
   (c) Terpenes
   (d) Lipoproteins

Amino Acids

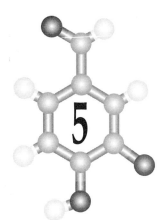

Amino acids are the essential components of all living cells. They are the building blocks of proteins. Although over 300 different amino acids are found in nature, only 20 amino acids are involved in proteins. Chemically, amino acids have a carboxyl group and and amino group, bonded to the same carbon atom. The general structural formula of the amino acids is:

$$H_2N - \underset{\underset{R}{|}}{\overset{\overset{H}{|}}{C}} - COOH$$

Amino acids differ from each other in R-group, which vary in structure, size and electric charge and influence the solubility of amino acids in water. In protein amino acids, it is always the α-carbon atom which is aminated and hence they are called α-*amino acids*. Since all the four groups linked to the α-carbon atom of amino acids are different (with the exception of glycine- for which R- is the hydrogen atom), the amino acids exist in two enantiomorphs–L-form and D-form. Amino acids, occurring in proteins belong to L-form.

5.1 ESSENTIAL AND NON-ESSENTIAL AMINO ACIDS

Most prokaryotic cells and some eukaryotic cells (e.g. plants, yeast) are capable of synthesizing all the amino acids present in the proteins. But higher animals, including man possess this ability for certain amino acids only. Those amino acids that can be synthesized from amphibolic intermediates are known as *nutritionally non-essential amino acids*, while those which cannot be synthesized in the body are known as *essential amino-acids* (Table 5.1). The essential amino acids must be obtained from the diet.

5.2 CLASSIFICATION OF AMINO ACIDS

On the basis of their R-groups, the amino acids can be classified into the following groups:

Table 5.1 Essential and non-essential amino acids

Amino Acid	Abbreviation/Symbol		Amino Acid	Abbreviation/Symbol	
Essential Amino Acids			**Non-essential Amino Acids**		
1. Arginine	Arg	R	1. Alanine	Ala	A
2. Histidine	His	H	2. Asparagine	Asn	N
3. Isoleucine	Ile	I	3. Aspartic Acid	Asp	D
4. Leucine	Leu	L	4. Glutamic Acid	Glu	E
5. Lysine	Lys	K	5. Cysteine	Cys	C
6. Methionine	Met	M	6. Glutamine	Gln	Q
7. Phenylalanine	Phe	F	7. Glycine	Gly	G
8. Threonine	Thr	T	8. Proline	Pro	P
9. Tryptophan	Trp	W	9. Serine	Ser	S
10. Valine	Val	V	10. Tyrosine	Tyr	Y

5.2.1   Amino Acids with Non-polar R-groups

The R-group in this class of amino acids are non-polar and hydrophobic. The following amino acids are classified as having non-polar side chains.

Glycine   Alanine   Valine

Leucine   Isoleucine

Methionine   Proline

Phenylalanine   *Tryptophan*

5.2.2 Amino Acids with Uncharged Polar Side Chains

The R-groups of these amino acids are more soluble in water, or more hydrophilic, than those of the non-polar amino acids, as they contain functional groups that form hydrogen bonds with water. The following are classified under this category:

Serine   *Threonine*

Asparagine   *Glutamine*

Tyrosine   *Cysteine*

5.2.3 Amino Acids with Charged Polar Side Chains

The charged side chains may be either positively charged or negatively charged. *Lysine, arginine* and *histidine* are amino acids with *positively charged side chains (basic)* at physiological pH values. Histidine is the only common amino acid having an ionizable side chain with pK_R near neutrality.

Lysine

$$
\begin{array}{c}
COO^- \\
| \\
H - C - CH_2 - CH_2 - CH_2 - NH - C \\
| \\
NH_3^+
\end{array}
\begin{array}{c}
NH_2 \\
\\
NH_2^+
\end{array}
$$

Arginine

$$
\begin{array}{c}
COO^- \\
| \\
H - C - CH_2 - \\
| \\
NH_3^+
\end{array}
$$

Histidine

The side chains of the acidic amino acids, *aspartic acid* and *glutamic acid* are negatively *charged* above pH 3. These amino acids are often called as *aspartate* and *glutamate* to emphasize that their side chains are usually negatively charged at physiological pH.

$$
\begin{array}{c}
COO^- \\
| \\
H - C - CH_2 - C \\
| \\
NH_3^+
\end{array}
\begin{array}{c}
O \\
\\
O^-
\end{array}
$$

Aspartic acid

$$
\begin{array}{c}
COO^- \\
| \\
H - C - CH_2 - CH_2 - C \\
| \\
NH_3^+
\end{array}
\begin{array}{c}
O \\
\\
O^-
\end{array}
$$

Glutamic acid

5.3 NON-STANDARD AMINO ACIDS

In addition to the 20 common amino acids, there are certain other amino acids which are either the constituents of proteins or biologically active peptides. In addition, many amino acids may independently play a variety of biological roles.

Some of these amino acids result from the specific modification of an amino acid residue after the polypeptide chain has been synthesized. Amino acid modifications include the simple addition of small chemical groups to certain amino acid side chains–*hydroxylation* (4-hydroxyproline), *methylation* (6-N-methyllysine), *carboxylation* (γ-carboxyglutamate), *acetylation* (ε-N-acetyllysine) and *phosphorylation* (O-phosphoserine). 4-hydroxyproline is found in collagen, a fibrous protein of connective tissue. 6-N-methyllysine is a constituent of myosin.

4-hydroxy proline

$$
\begin{array}{c}
COO^- \\
| \\
H - C - CH_2 - CH_2 - CH_2 - CH_2 - NH - CH_3 \\
| \\
NH_3^+
\end{array}
$$

6-N-Methyllysine

$$\underset{\text{Y-Carboxyglutamate}}{\text{H}-\overset{\overset{\text{COO}^-}{|}}{\underset{\underset{\text{NH}_3^+}{|}}{\text{C}}}-\text{CH}_2-\text{CH}\overset{\text{COO}^-}{\underset{\text{COO}^-}{}}}$$

$$\underset{\varepsilon\text{-N-Acetyllysine}}{\text{H}-\overset{\overset{\text{COO}^-}{|}}{\underset{\underset{\text{NH}_3^+}{|}}{\text{C}}}-\text{CH}_2-\text{CH}_2-\text{CH}_2-\text{CH}_2-\text{NH}-\overset{\overset{\text{O}}{||}}{\text{C}}-\text{CH}_3}$$

$$\underset{\text{O-Phosphoserine}}{\text{H}-\overset{\overset{\text{COO}^-}{|}}{\underset{\underset{\text{NH}_3^+}{|}}{\text{C}}}-\text{CH}_2-\text{O}-\text{PO}_3^{z-}}$$

The amino acid residues in protein molecules are exclusively L-stereoisomers. But in a few, generally small peptides, including some bacterial polypeptides and certain peptide antibiotics, D-amino acids residues are found. The presence of D-amino acids renders bacterial cell walls less susceptible to attack by the peptidase.

5.4 PROPERTIES OF AMINO ACIDS

5.4.1 Physical Properties

Amino acids are usually colourless crystalline solids with high melting points (usually more than 200°C). Amino acids are usually soluble in water. However, the solubility varies for different amino acids.

All amino acids (except glycine) are *optically active*. They exist in L- and D-isomeric forms.

Three amino acids – tryptophan, tyrosine and phenylalanine absorb ultraviolet light. This accounts for the characteristic strong absorbance of light by most proteins at a wavelength of 280 nm.

5.4.2 Chemical Properties

5.4.2.1 Amphoteric Nature

Amino acids in aqueous solutions are ionized and can act as acids or bases. Substances having this dual nature are *amphoteric substances* and are called *ampholytes*. Since amino acids contain both amino and carboxyl group, they are capable of both donating and accepting protons.

$$\text{R}-\overset{\overset{\text{H}}{|}}{\underset{\underset{\text{NH}_3^+}{|}}{\text{C}}}-\text{COO}^- \rightleftharpoons \text{R}-\overset{\overset{\text{H}}{|}}{\underset{\underset{\text{NH}_2}{|}}{\text{C}}}-\text{COO}^- + \text{H}^+$$

(donation of proton)

$$R - \underset{\underset{NH_3^+}{|}}{\overset{\overset{H}{|}}{C}} - COO^- + H^+ \rightleftharpoons R - \underset{\underset{NH_3^+}{|}}{\overset{\overset{H}{|}}{C}} - COOH$$

(accepting proton)

5.4.2.2 Zwitterion

At a certain pH (isoelectric pH), the amino acid molecules in aqueous solution carry both the positive and negative charges in equal amount and exist as Zwitterion (dipolar ion). At this point the net charge on it is zero. The zwitterion solutions of amino acids do not migrate to any pole in an electric field. If however, acid is added to the solution, they become positively charged ions and move to the cathode. Addition of alkali makes them negatively charged and they move to anode.

$$R - \underset{\underset{NH_3^+}{|}}{\overset{\overset{H}{|}}{C}} - COOH \rightleftharpoons R - \underset{\underset{NH_3^+}{|}}{\overset{\overset{H}{|}}{C}} - COO^- \rightleftharpoons R - \underset{\underset{NH_2}{|}}{\overset{\overset{H}{|}}{C}} - COO^-$$

| *Basic form* | *Zwitterion* | *Acidic form* |

5.4.2.3 Peptide Bond Formation

Amino acids are the building blocks of proteins. The amino acids are joined together in the protein molecule by peptide bonds (—CO—NH—) formed by the condensation of α-COOH of one amino acid with α-NH_2 group of another one with the release of one molecule of water.

$$NH_2 - CH_2 - CO\,OH \qquad H\,NH - CH - COOH$$

Glycine $\qquad\qquad\qquad\qquad\qquad$ $\underset{CH_3}{|}$ $\quad$ Alanine

$$NH_2 - CH_2 - CO - NH - \underset{\underset{CH_3}{|}}{CH} - COOH \quad + H_2O$$

5.4.2.4 Ninhydrin Reaction

Amino acids yield coloured products with Ninhydrin. This reaction is most widely used for quantitative estimation of amino acids. Ninhydrin oxidatively decarboxylates α-amino acids to CO_2, NH_3 and an aldehyde with one carbon atom less than the parent amino acid. The reduced ninhydrin then reacts with ammonia and some more ninhydrin to form blue-violet compounds.

Coloured product

The intensity of the colour is determined colorimetrically, which is proportional to the amount of amino acid present. This method is used extensively in the determination of amino acids.

5.4.3 General Reactions

Amino acids give all the typical chemical reactions associated with compounds that contain carboxyl and amino groups.

5.4.3.1 Reactions due to Carboxyl (—COOH) group

(i) **Formation of esters:** Amino acids can form esters with alcohols. The —COOH group can be esterified with alcohol.

$$\underset{\text{Glycinehydrochloride}}{\overset{\displaystyle COOH}{\underset{\displaystyle CH_2-NH_3Cl}{|}}+C_2H_5OH} \longrightarrow \underset{\text{Esterhdyrochloride}}{\overset{\displaystyle COOC_2H_5}{\underset{\displaystyle CH_2-NH_3Cl}{|}}+H_2O}$$

(ii) Formation of amines: Action of specific amino acid decarboxylases, dry distillation or heating with Ba $(OH)_2$ evolves CO_2 from —COOH group and changes the amino acid into its amine.

$$\overset{\displaystyle NH_2}{\underset{\displaystyle COOH}{\underset{|}{R-CH}}} \xrightarrow[\text{Heat}]{Ba(OH)_2} CO_2\uparrow + \underset{\text{Amine}}{\underline{R-CH_2-NH_2}}$$

5.4.3.2 Reactions due to amino (—NH₂) group

(i) Salt formation with acids: The basic amino acids react with mineral acids to form salts like hydrochlorides.

$$\underset{\text{Glycine}}{\overset{\displaystyle COOH}{\underset{\displaystyle H_2C-NH_2}{|}}+HCl} \longrightarrow \underset{\text{Glycine hydrochloride}}{\overset{\displaystyle COOH}{\underset{\displaystyle H_2C-NH_3Cl}{|}}}$$

(ii) Reaction with HNO₂: The amino acids except proline and hydoxyline react with HNO_2 (nitrous acid) and librate N_2 from NH_2 group.

$$\overset{\displaystyle NH_2}{\underset{\displaystyle H}{\underset{|}{R-C-COOH}}} \xrightarrow{HNO_2} \underset{\alpha\text{-hydroxy acid}}{\overset{\displaystyle OH}{\underset{\displaystyle H}{\underset{|}{R-C-COOH}}}+N_2\uparrow+H_2O}$$

This forms the basis for Van Slyke method for determining amino nitrogen in amino acids and proteins.

5.5 SUMMARY

- Amino acids are the building blocks of proteins. Chemically, they have a carboxyl group and an amino group bonded to the same carbon atom.
- Amino acids which can be synthesized in the body are called non-essential amino acids. Essential amino acids are those which cannot be synthesized in the body of the animals.
- The R-group present in the amino acids may be either polar or non-polar.
- Amino acids are colourless, crystalline solids with high melting points.
- In aquous solutions, amino acids are ionized and act as ampholytes.

- Amino acids give all the typical reactions associated with compounds containing carboxyl and amino groups.

EXERCISE

1. What are amino acids? Describe the physical and chemical properties of amino acids.
2. Differentiate between essential and non-essential amino acids.
3. Write notes on:
    (a) Zwitterion
    (b) Non-standard amino acids

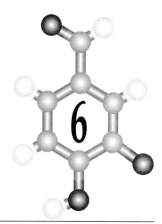

Proteins

Proteins are the high molecular weight polypeptides. The term protein was first proposed by Swedish chemist Berzelius and derived from the Greek word 'protios' meaning 'holding the first place'. Gerardus Mulder, for the first time, used the term and referred it to the complex organic nitrogenous substances found in the cells of living organisms. Proteins are the essential constituents of living cells. All the basic functions of life depend upon specific proteins.

Proteins perform a wide variety of functions. Depending upon their chemical and physical structures, they are involved in catalysis, contraction, conduction, nutrition, binding, defense or structure.

6.1 CHEMICAL STRUCTURE

Proteins are the polymers of amino acids. In proteins the amino acids are linked together by peptide bonds. The carboxyl group of one amino acid is joined to the amino group of another acid to form a peptide bond with the release of a water molecule.

$$R_1 - \overset{\overset{\displaystyle NH_2}{|}}{\underset{\underset{\displaystyle H}{|}}{C}} - \overset{\overset{\displaystyle O}{\|}}{C} - OH + H - \overset{\overset{\displaystyle H}{|}}{N} - \overset{\overset{\displaystyle H}{|}}{\underset{\underset{\displaystyle R_2}{|}}{C}} - COOH$$

$$\searrow H_2O$$

The product formed by a peptide bond is a peptide. When a large number of amino acids join together they form polypeptide chains. The amino acid molecules in a polypeptide chain are linked by peptide bonds.

$$R_1 - \overset{\overset{\displaystyle NH_2}{|}}{\underset{\underset{\displaystyle H}{|}}{C}} - \overset{\overset{\displaystyle O}{\|}}{C} - \overset{\overset{\displaystyle H}{|}}{N} - \overset{\overset{\displaystyle H}{|}}{\underset{\underset{\displaystyle R_2}{|}}{C}} - \overset{\overset{\displaystyle O}{\|}}{C} - \overset{\overset{\displaystyle H}{|}}{N} - \overset{\overset{\displaystyle H}{|}}{\underset{\underset{\displaystyle R_3}{|}}{C}} - \overset{\overset{\displaystyle O}{\|}}{C} - OH$$

Peptide bond

The polypeptide chains form proteins. A protein may consist of a single or more polypeptide chains e.g. myoglobin consists of a single polypeptide chain while haemoglobin consists of four polypeptide chains.

6.2 BIOLOGICAL SIGNIFICANCE

Proteins are the most versatile macromolecules in the living systems and play crucial roles in all biological processes.

Structural proteins are involved in the formation and maintenance of various cellular structures. They are the part of cell wall, cell membrane and primary fibrous constituents of the cells. Collagen is the most abundant animal protein which form a major part of the skin, cartilage, ligament, tendons and bones. Keratin is involved in the formation of derivatives of skin such as hair, feathers, horns, hoofs and claws. Muscle proteins-actin and myosin are involved in the process of contraction.

Almost all chemical reactions in the organism are catalyzed by enzymes which are protein in nature. Some enzymes are simple proteins containing only amino acids while others are complex enzymes which contain some other molecule also, in addition to proteins.

Some proteins take part in the transport of material. Haemoglobin is an important protein involved in the transport of gases. Myoglobin transports oxygen in the muscles. The membrane proteins transport ions and small molecules across the cell membrane.

Certain proteins act as storage proteins. In the liver ferritin stores iron. Seeds store nutritional proteins.

An important function of the proteins is in immune system. They are the important part of immune system. Antibodies are immunoglobulin which neutralize the antigens entering the body. Some hormones are also protein in nature.

In addition, proteins also take part in blood coagulation. Fibrinogen and thrombin are important proteins involved in blood coagulation.

6.3 CLASSIFICATION

Proteins can be divided into different types according to their *functions, shape, solubility, physical state* etc.

On the basis of their conformation the proteins can be grouped into two major classes:
1. Fibrous Proteins
2. Globular Proteins

6.3.1 Fibrous Proteins

In fibrous proteins polypeptide chains are arranged in parallel along a single axis to yield long fibres or sheets. These are insoluble in water or dilute salt solutions. These proteins are the basic structural elements in animal tissues. *Collagen* of tendons and bone matrix, *keratin* of hair and skin and *elastin* all are fibrous proteins.

6.3.2 Globular Proteins

In globular proteins the polypeptide chains are tightly folded into compact spherical or globular shapes. Most globular proteins are soluble in water. All enzymes, certain hormones, antibodies are the examples of globular proteins.

On the basis of their structure and complexity proteins are classified into three major groups:

1. Simple proteins
2. Conjugated or complex proteins
3. Derived proteins

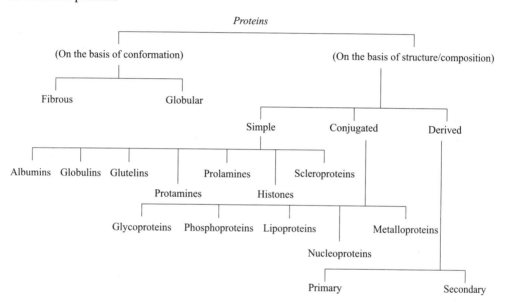

6.3.3 Simple Proteins

They are the proteins which consist solely of amino acids. They are further classified into the following subgroups:

- **Albumins:** Albumins are soluble in water and coagulate on heating. They are precipitated by saturated solution of ammonium sulphate *e.g. serum albumin, egg albumin, lactalbumin (milk), legumelin (legumes), leucosin(cereals).*
- **Globulins:** They are insoluble in water but soluble in dilute neutral salt solutions. They coagulate on heating. They are precipitated by half saturation with ammonium sulphate *e.g. plasma globulin, ova globulin, serum globulin.*
- **Glutelins:** These are insoluble in water but are soluble in dilute acids and bases. On heating these coagulate e.g. *glutenin* in wheat and *oryzenin* in rice.
- **Prolamines:** These are insoluble in water, absolute alcohol and other neutral solvents but dissolve in 50-80% ethanol. They are not coagulated by heat e.g. *zein* (maize), *gliadin* (wheat).
- **Scleroproteins:** These are also known as albuminoids. They are soluble in water and solutions of neutral salts e.g. *keratin, collogen, elastin.*

- **Histones:** Histones are soluble in water and dilute acids but insoluble in ammonia. They are not coagulated by heat e.g. *nucleoproteins.*
- **Protamines:** These are basic proteins which are soluble in water, dilute acids and alkalis and ammonium hydroxide. They are not coagulated by heat e.g. *salmine, sturine*

6.3.4 Conjugated Proteins

These are the proteins which consist of not only amino acids but also some organic or inorganic components. The non-amino acid substance linked to proteins is called prosthetic group. On the basis of their prosthetic group conjugated proteins are of the following types:

- **Glycoproteins:** Glycoproteins are the proteins with covalently linked carbohydrate group. The carbohydrate in glycoproteins is either a monosaccharide or short oligosaccharide. The percent by weight of carbohydrate groups in different glycoproteins may vary from less than 1% in ova albumin to as high as 80% in mucoproteins. Glycoproteins which have a very high carbohydrate content are called *proteoglycans.*

  The glycoproteins may contain from two to dozens of monosaccharide residues usually of two or more kinds. The oligosaccharide groups of most glycoproteins are covalently attached to the R-groups of specific amino residues in polypeptide chains. They are present both in animals and plants but not in bacteria e.g. egg albumin, serum albumin.

- **Phosphoproteins:** In phosphoproteins, the protein is attached to the phosphoric acid. The phosphoric acid is attached to the hydroxyl group of protein by an ester linkage e.g. casein and vitelline.

- **Lipoproteins:** In these proteins phospholipids/cholesterols are attached as prosthetic group e.g. lipoproteins are present in blood, milk, egg yolk etc.

- **Nucleoproteins:** Nucleoproteins contain nucleic acids as the prosthetic group e.g. nucleohistones.

- **Metalloproteins:** In metalloproteins, proteins are linked to metallic prosthetic groups. The metallic group gives colour to the proteins. These are also known as *chromoproteins* e.g. haemoglobin, hemocynin etc.

6.3.5 Derived Proteins

These are the intermediate products formed during hydrolysis of proteins. They also include the products of decomposition of protein molecule and isolated protein molecules after the removal of the prosthetic groups from the conjugated protein. Derived proteins can be divided into two major groups:

1. **Primary derived proteins**
2. **Secondary derived proteins**

Primary derived proteins: These are denatured or coagulated proteins. The denaturation is caused by heat, acid or alkali treatment. Their molecular weight is same as the native protein but they differ in solubility, precipitation and crystallization. They can be of the following types:

(i) **Proteans:**   They are the first products produced by the action of acids, enzymes or water on proteins. They are insoluble in water e.g. *myosan* derived from mysin, *fibrin* derived from fibrinogen and *edestan* derived from edestin.

(ii) Metaproteins: They are formed from further action of acids and allkalies on proteins. They are insoluble in water but generally soluble in dilute acids and alkalies e.g. acid and alkali metaproteins such as alkali and acid albuminates.

(iii) Coagulated proteins: These are produced by the action of heat or alcohol on proteins e.g. coagulated egg albumin.

Secondary derived proteins: These are formed by the progressive hydrolysis of proteins at their peptide linkage. They are mainly of three types i.e. proteoses, peptones and peptides.

(i) Proteoses: Proteoses are the hydrolytic products of proteins and are soluble in water. These are precipitated by saturating their solutions with ammonium sulphate.

(ii) Peptones: Peptones are the hydrolytic products of proteoses and are soluble in water. They are not precipitated by saturation with ammonium sulphate but can be precipitated by phosphotungstic acid.

(iii) Peptides: Peptides are the protein derivative amino acid units joined as peptide bonds. They are also soluble in water and can be precipitated by phosphotungstic acid.

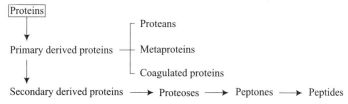

6.4 GENERAL PROPERTIES

Proteins are normally colourless (expect chromoproteins), tasteless and ordourless substances. The solubility of proteins varies according to their nature. Some proteins are soluble in water, acid and salt solutions while they are insoluble in alcohol or ether.

Like amino acids proteins are amphoteric in nature. This is due to the presence of several free —NH$_2$ and —COOH groups.

Proteins are either positively or negatively charged molecules and in an electric field migrate either towards cathode or towards anode. At isoelectric pH they are electrically neutral and do not move towards any pole. The isoelectric pH of the proteins has been found to depend upon the relative number of acidic or basic groups which are released by amino acids.

Table 6.1 Isoelectric pH of some common proteins

Protein	pH
Pepsin	2.7
Casein	4.6
Avidin	10
Serum globulin	5.4
Gliadin	9
Serum albumin	4.7
Cytochrome	9.8

Hydrolysis of proteins: The proteins undergo hydrolysis by acids, alkali or hydrolytic enzymes. The hydrolysis ultimately leads the proteins to amino acids. Complete hydrolysis with HCl or H_2SO_4 yields free amino acids and their breakdown products.

Denaturation of proteins: When proteins are treated with heat, UV rays, X-rays, high pressure, certain organic solvents, acids or alkali or detergents etc., they undergo remarkable changes in their solubility, optical rotation and biological properties. These changes occuring in proteins are collectively called *denaturation*. Denaturation is the result of changes in conformation or unfolding of protein molecules. But there may not be a complete unfolding of the proteins. Under most of the conditions, denatured proteins exist in folded states. There is a disruption of secondary, tertiary and quaternary organization of a protein molecule due to cleavage of mono-covalent bonds but the primary structure of proteins is not affected. In denatured proteins the solubility is decreased or lost. Proteins can also be denatured by extreme pH which alters the net charges on the protein causing electrostatic repulsion and the disruption of some hydrogen bonding. Denaturation destroys enzymal and hormonal activity and the proteins become biologically inactive.

The process of denaturation in some proteins is reversible. If the denatured protein is returned to the conditions in which their native conformation is stable, they regain their native structure and biological activity. This process is called **renaturation** (Fig. 6.1).

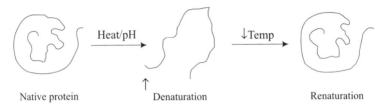

Fig. 6.1   Denaturation and renaturation of proteins

6.5   STRUCTURAL ORGANIZATION OF PROTEINS

Proteins are linear polymers formed by linking of one amino acid to another with a peptide bond. The peptide chain has a variable length depending on the number of amino acids linked. In proteins four structural levels are present (Fig. 6.2). They are:

- Primary Structure
- Secondary Structure
- Tertiary Structure
- Quaternary Structure

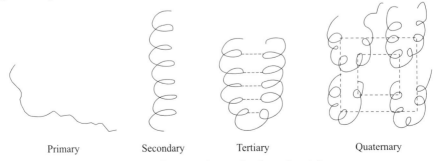

Fig. 6.2   Structural organization of proteins

6.5.1 Primary Structure

Primary structure refers to the linear sequence of amino acid residues of a protein. Each protein has a unique, precisely defined amino acid sequence. The amino acid sequence of proteins is genetically determined which depends on DNA which specifies the nucleotide sequence of mRNA which in turn specifies the amino acid sequence of a protein. In a protein the α-carboxyl group of one amino acid is linked to the α-amino group of another amino acid with a peptide bond accompanied by the loss of a water molecule. A series of amino acids joined by peptide bonds form a polypeptide chain. The number of amino acids in a polypeptide chain is varible. If the number of amino acids is low then it is called a oligopeptides or simply peptides. Each polypeptide chain has at one end N-terminal amino acid containing a free amino group and at the other end a c-terminal amino acid containing free carboxyl group. By convention, the amino acid end (N-terminal) is taken to be the beginning of a polypeptide chain and the sequence of amino acids in a polypeptide chain is written starting with the amino terminal residue.

The sequence of amino acids in a protein is very important. The amino acid sequence determines the three dimensional structure of proteins and is essential to elucidating its mechanism of action. Any alteration in amino acid sequence can produce a protein with abnormal function and diseases e.g. sickle cell anemia can result from a change in a single amino acid.

6.5.2 Secondary Structure

Secondary structure refers to regular folding patterns of contiguous portions of the polypeptide chain. The secondary structure is derived from the primary structure by the formation of hydrogen bond interaction between amino acid residues close to one another which may lead to the folding of the polypeptide chain. In 1951 Linus Pauling and Robert Corey proposed two periodic structures:

   (i) α helix, and
   (ii) β-pleated sheet

α-helix: α-helix is a rod like structure in which a tightly coiled backbone forms the inner part of the rod and the side chains extend outward in a helical array. The helical structure of protein is formed by the hydrogen bonds between peptide groups within the same polypeptide chain. A hydrogen bond is formed between the CO-group of each amino acid with the –NH group of the amino acid situated four residue ahead in the sequence. The regular appearance of hydrogen bonds between every first and fourth peptide group determines the regularity of turns in

Fig. 6.3 Amino acid sequence of bovine insulin

polypeptide chain. In α-helix each residue is related to the next one by a rise of 1.5 A° along the helix and a rotation of 100 degrees. Thus, there are 3.6 amino acid residues per turn of helix. All α-helix found in proteins are right handed. Myosin and tropomyosin in muscles, keratin in hair and fibrin in blood clots show α-helix structure.

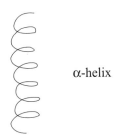

α-helix

Fig. 6.4 α-helix structure of protein

β-pleated sheet: β-pleated structure was proposed by Pauling and Corey. The β-pleated sheet differs markedly from the α-helix. In β-pleated sheets, hydrogen bonding occurs between neighbouring polypeptide chains rather than within one as in a α-helix. β-pleated sheet structure is of two types–parallel and antiparallel (Fig. 6.5).

In parallel β-pleated sheets the N-terminal end of the polypeptide chains point in the same direction i.e. the chain extend in the same direction e.g. keratin. In antiparallel β-sheets, the neighbouring hydrogen bonded polypeptide chains run in opposite directions. e.g. silk fibroin. β-sheets contain 2 to >12 polypeptide strands. Out of the two types of β-sheets parellel β-sheets are less stable than antiparallel β-sheets possibly because the hydrogen bonds of parallel sheets are distorted as compared to those of antiparallel sheets.

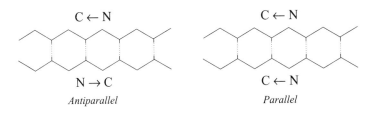

Fig. 6.5 β-pleated structure of protein

6.5.3 Tertiary Structure

Tertiary structure refers to the three dimensional structure, especially the bonds between amino acid residues that are distant from each other in the polypeptide chain and the arrangements of secondary structure elements relative to one another. Tertiary structure is more complex than the secondary structure and is found in globular proteins. X-ray crystallography and nuclear magnetic resonance (NMR) studies have revealed the three dimensional structure of several proteins. Myoglobin, ribonuclease, cytochrome C exist in tertiary structure.

Myoglobin is a single polypeptide chain of 153 amino acids containing a heme (iron-containing porphyrin) group in the center (Fig. 6.6). The molecular weight is 16,700 Daltons. It is the primary oxygen carrying pigment of muscle tissues. In 1957, John Kendrew and associates successfully determined the structure of myoglobin by high resolution X-ray crystallography. Myglobin is an extremely compact molecule with overall dimensions $45 \times 35 \times 25$A°. A myoglobin polypeptide is comprised of 8 separate right handed α-helices, designated A through H, that are connected by short non-helical regions.

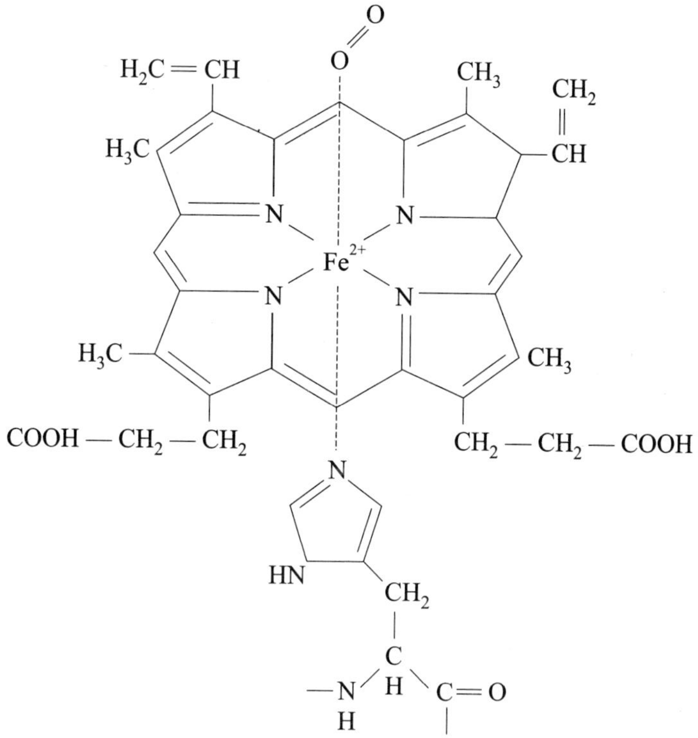

Fig. 6.6 The heme group

Each myoglobin molecule contains one heme prosthetic group inserted into a hydrophobic cleft in the protein. The heterocyclic ring system of heme is a porphyrin derivative containing four pyrrole groups linked by methane bridges. Each heme residue contains one central coordinately bound iron atom that is normally in the Fe^{2+}. The Fe^{2+} atom at the center of the heme is coordinated by four porphyrin N atoms and one N from a His side chain. It is the oxygen carrier in muscles and the oxygen carrying (binding) capacity depends on the presence of 'heme'. When exposed to oxygen, the Fe^{2+} atom of the isolated heme is irreversibly oxidized to Fe^{3+}. The protein portion of myoglobin prevents this oxidation and makes it possible for O_2 to bind reversibly to the heme group.

6.5.4   Quaternary Structure

Most proteins with molecular masses > 100 KD consist of more than one polypeptide chain. Each polypeptide chain in such a protein is called a subunit. The subunits associate with a specific geometry. Quaternary structure refers to the spatial arrangement of these subunits. In the simplest quaternary structure the protein consists of two subunits e.g. enzyme *phosphorylase a* contains two subunits which are identical to each other and alone each of them is catalytically inactive. When they join together they form a dimmer, an active form of the enzyme. In this case where the subunits are identical to each other, it is called *homogeneous quaternary structure*. When these subunits are non-identical then it is called a *hetergeneous quaternary structure*. Human haemoglobin, for example, consists of four subunits – two subunits of one type-α and two subunits of another type-β.

Haemoglobin is a respiratory pigment present in the blood of most of the animals. It is present in the red blood corpuscles of the vertebrates and in dissolved condition in the plasma of the blood of some invertebrates. The structure of haemoglobin was determined by Max Perutz. It is a chromo protein consisting of two parts-one part is a specific protein known as globin (96%) and the other part is a non-specific prosthetic group- the iron containing heme (Fig. 6.7). Heme is a metalloprotein, contaning iron as the metal. In haemoglobin iron remains in ferrous Fe^{2+} form. Globin helps heme to keep the iron in Fe^{2+} form and to combine loosely and reversibly with oxygen. Mammalian haemoglobin is a tetrameric protein with the quaternary structure $\alpha_2\beta_2$. Normal adult haemoglobin consists of two α chains and two β chains while the fetal haemoglobin consists of two α and two γ chains. The α chains are made up of 141 amino acids and β chains and γ chains are made up of 146 amino acids. In globin the four peptide chains are believed to held together by a non-covalent factor. The molecular weight of the molecule is 68000 Daltons. The haemoglobin molecule has overall dimensions $64 \times 55 \times 50A°$. The most characteristic property of haemoglobin is its affinity with O_2 and CO_2. It combines with O_2 and forms oxyhaemoglobin. The structure of haemoglobin and oxyhaemoglobin are noticeably different from each other.

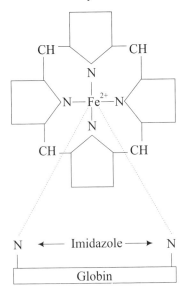

Fig. 6.7   Haemoglobin

Proteins with more than one subunits are called oligomers and their subunits are called promoters. In most of the oligometric proteins the promoters are symmetrically arranged i.e. each promoter occupies a geometrically equivalent position in the oligomers.

6.6   SUMMARY

- Proteins are the polymers of amino acids in which amino acids are linked together by peptide bonds.
- Proteins are the building blocks of the body and they are involved in a variety of functions in the living body. They function as catalyst; provide mechanical support and immune protection and control growth and differentiation.
- Proteins can be classified on the basis of their conformation, solubility, structure and complexity.
- Protein structure can be described at four levels. Primary structure refers to the amino acid sequence. Secondary structure refers to the regular polypeptide folding pattern such as helices and β-sheets. Tertiary structure describes the folding of secondary structural elements of the proteins. Proteins with more than one polypeptide chains exhibit quaternary structure which refers to the spatial arrangement of the subunits in a protein.
- The protein can be denatured by variety of conditions and substances such as heat, pH, detergents and urea etc.

EXERCISE

1. What are proteins? Explain their biological significance.
2. Classify proteins on the basis of their structure and complexity.
3. Describe the helix and pleated sheet structures of proteins.
4. Write short notes on:
   (a) Denaturation of proteins
   (b) Structure of haemoglobin
   (c) Glycoproteins

Enzymes

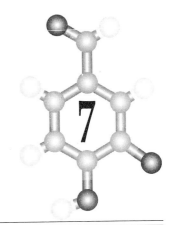

In all living organisms some specific molecules are required to catalyze the chemical reactions occuring in the body. Such molecules are known as enzymes. Enzymes are the biocatalysts, which like chemical catalysts are not used up in the reactions but unlike catalyst they are produced by living cells. The term enzyme was first used by Frederick W. Kuhne. In 1926 James Sumner isolated and crystallized the enzyme urease.

Nearly all enzymes are proteins (exception of a small group of catalytic RNA molecules). Enzymes accelerate the rate of reactions by a million or more times. They are highly specific in nature. Their specificities are both in the reactions that they catalyze and their substrates.

Some enzymes require no chemical groups for their activity other than their amino acid residues. They are simple proteins. Such enzymes are called *Simple enzymes*. e.g. amylase, urease etc. There are other enzymes which require presence of an additional chemical component called a *cofactor* for their activity. Such enzymes are called *Conjugated* or *Holoenzymes*. Thus a holoenzyme consists of two parts:

$$\text{Apoenzyme} \quad + \quad \text{Cofactor} \quad = \quad \text{Holoenzyme}$$
$$\uparrow \qquad\qquad\qquad \uparrow$$
$$\text{Protein part} \quad \text{Non-protein part}$$

The cofactor in a holoenzyme may be one or more inorganic ion, metal ion or a complex organic molecule. If the cofactor is a complex organic molecule then it is called a *coenzyme*. If the coenzyme or metal ion is tightly or even covalently bound to the enzyme protein, it is called a *prosthetic group*. e.g. the heme prosthetic group of cytochrome C is tightly bound to the protein through extensive hydrophobic and hydrogen bonding interactions together with covalent bonds between the heme and specific protein side chains.

7.1 CLASSIFICATION OF ENZYMES

There are various ways of naming and classifying the enzymes. Many enzymes are named by adding the suffix-'ase' to the name of their substrates e.g. enzyme catalyzing the hydrolysis of urea is urease, enzyme acting on lipids are named as lipases. Certain enzymes are named by adding the suffix 'ase' to the reaction they catalyze e.g. the enzyme catalyzing hydrolysis is named as hydrolase; the enzyme which brings about oxidation is named as oxidase. Similarly, the suffix- lytic is used to denote enzyme splitting the substrates such as *proteolytic*-protein splitting; *lipolytic-* fat splitting etc.

In 1961, the commission on enzymes of the International Union of Biochemistry (IUB) framed certain rules for the nomenclature and classification of enzymes. According to this system, there are six classes of enzymes which are further divided into subclasses and sub-subclasses, based on the type of reaction catalyzed. Each enzyme is assigned a four part classification number and a systematic name. For example, enzyme carboxypeptidase A has the systematic name *peptidyl-L-amino acid hydrolase* and the classification number EC 3.4.17.1. Here E.C. stands for Enzyme Commission and the numbers represent the class, subclass, sub-subclasses and its arbitrarily assigned serial number in its sub-subclass.

Table 7.1 Classification of enzymes according to IUB

Class	Type of Reaction Catalysed
1. Oxidoreductases	Oxidation-reduction reaction
2. Transferases	Transfer of functional groups
3. Hydrolases	Hydrolysis reactions
4. Lyases	Group elimination to form double bonds
5. Isomerases	Isomerisation
6. Ligases	Bond formation coupled with ATP hydrolysis

According to IUB classification, six main classes of enzymes are as follows:

(i) *Oxidoreductases:* The enzymes which catalyze oxidation and reduction reactions are called oxidoreductases. This class includes dehydrogenase, oxidase, hydroxylase, catalases enzymes.

(ii) *Transferases:* The enzymes of this class catalyze the transfer of functional groups between a donor and an acceptor molecule e.g. *amino transferases* transfer amino group from one amino acid to another; *glycosyl transferases* catalyse the transfer of an activated glycosyl residue to a glycogen primer.

(iii) *Hydrolases:* Hydrolases are the enzymes which catalyze the hydrolysis of their substrates by adding water across the bond they split e.g. lipases, proteases, glycosidases etc.

(iv) *Lyases:* The enzymes which catalyze the removal of groups from substrate by mechanisms other than hydrolysis leaving double bonds are included in this class e.g. aldolase, enolase, fumarase etc.

(v) *Isomerase:* Isomerase are those enzymes which catalyze interconversion of isomers by intramolecular arrangements of atoms or groups e.g. epimerases, glucose-phosphate isomerase etc.

(vi) *Ligases:* These enzymes catalyze the linking of two compounds using the energy released from the hydrolysis of ATP. These enzymes are also called synthetases e.g. enzyme *glutamine synthetase* catalyzes the formation of glutamine from glutamate and NH_3 using ATP.

7.2 GENERAL PROPERTIES

All enzymes are proteinous, in nature. They may be either simple or conjugated proteins. Enzymes are required in very small amount. A single enzyme can act upon thousands of substrate molecules per minute. The number of substrate molecules catalyzed by an enzyme is called *turn over number*. Enzymes are very specific in their action. Each enzyme acts on a particular substrate or a group of related substrates. This property of enzymes is called *specificity* of enzymes. Thus, the specificity of an enzyme may be *absolute specificity* (when enzyme acts on a particular substrate e.g. lactase acts on lactose), *group specificity* (when an enzyme acts on a particular group e.g. trypsin acts on peptide bonds) or *optical specificity* (when an enzyme acts on a particular isomers e.g. D-amino acid oxidase acts on D-amino acids).

Enzymes are very sensitive to temperature. Each enzyme has an optimal temperature at which the rate of activity is maximum. For most enzymes, the optimal temperature falls between 30° to 40°C. The rate of reaction decreases when the temperature goes up or goes down from the optimal temperature. At very high temperature enzymes, being protein, denature.

Enzymes are biocatalysts and like chemical catalysts participate in the biological reactions but itself remain unchanged at the end of the reaction.

The activity of enzymes also depends on the pH of the medium. Each enzymes acts on a particular pH, either acidic, or basic medium. The effectiveness of an enzyme varies greatly with change of pH.

7.3 MECHANISM OF ENZYME ACTION

Michaelis and Menten proposed a hypothesis for enzyme action. According to this hypothesis, in enzyme catalyzed reaction, first the enzyme (E) combines with the substrate (S) molecule to form enzyme–substrate complex (ES). Enzyme-substrate complex is a transtitional state. This complex is then dissociated to form product (P) and enzyme (E).

$$E + S \rightleftharpoons ES \rightarrow P + E$$

The binding of the enzyme with substrate is highly specific in nature. The substrate molecule binds with its enzyme at a specific site called as *active site* or *catalytic site*. The active site is made up of several amino acid residues and possesses a complex three dimensional form and shape. The binding of enzyme to its substrate is a physical process. To explain the formation of enzyme-substrate complex, two mechanisms have been proposed:

1. Lock and Key model
2. Induced Fit model

7.3.1 Lock and Key Model

Lock and Key model was proposed by *Emil Fischer*. According to this model a substrate binds with its enzyme at specific site known as active site. The active site already exists in the enzyme. The molecular shape of the enzyme surface is such that only a particular substrate molecule can be adjusted in it to form enzyme-substrate complex. Thus, a substrate fits in the enzyme in the same manner as a key fits in the lock. The active site provides a rigid, pre-shaped template for substrate molecule (Fig. 7.1).

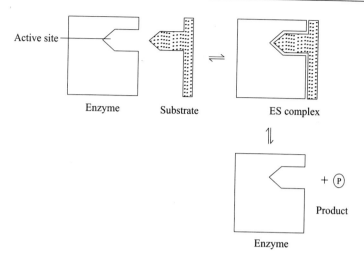

Fig. 7.1 Lock-and key model

7.3.2 Induced-fit Model

Induced fit model was proposed by *Koshland.* An essential feature of this model is the flexibility of the active site. According to this model when an enzyme binds with a substrate, the substrate induces a conformational change in the enzyme. This aligns amino acid residues or other groups on the enzyme in the correct spatial orientation for proper binding of the substrate. Thus, the active site does not possess a rigid, preformed structure on the enzyme but substrate can induce some change in it so that it (substrate) can fit properly (Fig. 7.2).

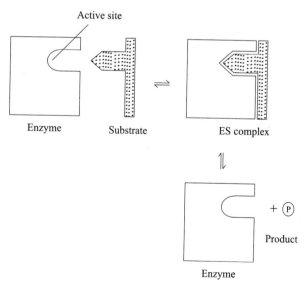

Fig. 7.2 Induced-fit model

Induced fit model differs from the lock and key model in that, in lock and key model the active site is presumed to be pre-shaped to fit the substrate. In induced fit model the substrate induces a conformational change in the enzyme. This model has gained much experimental support.

7.3.3 Kinetics of Enzyme Action

The rate of enzyme action depends on the relative concentrations of enzyme and its substrate. If the concentration of enzyme is large, the rate of reaction increases with the increase in the concentration of substrate. There is a linear relationship between the substrate concentration and the rate of reaction up to a level. The initial velocity increases almost linearly with an increase in the substrate concentration but beyond a level, velocity increases but not in a linear fashion. Finally, a point is reached beyond which any further increase in the concentration of substrate does not increase the rate of reaction. This is the maximum velocity (V_{max}). The rate of reaction is limited by the fact that when substrate concentration reaches a specific level, it utilizes all the available groups of enzymes. Beyond this level, no more active enzyme group is available, so the velocity is maximum and constant. Any further increase in the concentration of substrate will diminish the velocity of reaction. The behaviour of enzymes with substrate concentration was studied by *Leonor Michaelis* and *Maud Menten*.

In an enzyme reaction, it is postulated that first the enzyme combines with the substrate to form enzyme-substrate complex. This enzyme substrate complex breaks down to form product and enzyme is released.

$$E + S \underset{k_2}{\overset{k_1}{\rightleftharpoons}} ES \xrightarrow{k_3} E + P \qquad \qquad ...(1)$$

The rate limiting step is the breakdown of ES to product and free enzyme.

The rate of reaction is determined by the breakdown of ES to form product:

$$V = k_3 [ES] \qquad \qquad ...(2)$$

At any moment rate of formation of ES depends on the molar concentration of enzyme, substrate and the enzyme substrate complex:

$$V_1 = k_1 \{[E] - [ES]\}[S]$$

Rate of disappearance of ES depends on its dissociation into E and S and its decomposition into enzyme E and product P:

$$V_2 = k_2 [ES] + k_3 [ES]$$

If we consider a steady state, where the rate of formation of ES and rate of break down of ES are equal to each other, i.e.

$$V_1 = V_2$$

$$K_1 \{[E] - [ES]\}[S] = k_2 [ES] + k_3 [ES]$$

or, $\qquad k_1 [E][S] - k_1 [ES][S] = [k_2 + k_3][ES]$

$$k_1 [E][S] - k_1 [ES][S] + k_1 [ES][S]$$

$$= [k_2 + k_3][ES] + k_1 [ES][S]$$

$$k_1 [E][S] = \{k_1 [S] + k_2 + k_3\}[ES]$$

$$[ES] = \frac{k_1[E][S]}{k_1[S] + k_2 + k_3}$$

or,

$$[ES] = \frac{[E][S]}{[S] + (k_2 + k_3)/k_1}$$

Here, $(k_2 + k_3)/k_1$ is defined as *Michaelis Constant K_m*.

We can write the above reaction as,

$$[ES] = \frac{[E][S]}{[S] + K_m} \qquad ...(3)$$

From equation (2),

$$[ES] = \frac{V}{k_3}$$

Thus, equation (3) can be written as

$$\frac{V}{k_3} = \frac{[E][S]}{[S] + K_m}$$

or,

$$V = \frac{k_3[E][S]}{[S] + K_m} \qquad ...(4)$$

At V_{max}, the enzyme is saturated with the substrate i.e. [ES] = [E].

When we consider velocity to be V_{max},

$$V = V_{max} = k_3 [ES] = k_3 [E]$$

The equation (4) becomes,

$$V = \frac{V_{max} [S]}{[S] + K_m}$$

or,

$$\boxed{V = \frac{V_{max}}{K_m}}$$

This is known as Michaelis-Menter equation. If we consider a condition, when $V = 1/2\ V_{max}$ then,

$$\frac{V_{max}}{2} = \frac{V_{max}[S]}{[S]+K_m}$$

$$\frac{1}{2} = \frac{[S]}{[S]+K_m}$$

or, $K_m = [S]$ when $V = \frac{1}{2}V_{max}.$

Thus Michaelis Constant K_m can be defined as the amount of the substrate concentration that produces half maximal velocity. It indicates the affinity of an enzyme for its substrate. A high K_m indicates a weak affinity of the enzyme to the substrate and a low K_m indicates strong affinity. The K_m value of different enzymes is highly variable and is a specific property of an enzyme. Isoenzymes have different K_m values. The K_m value is determined by plotting V against [S] (Fig. 7.3).

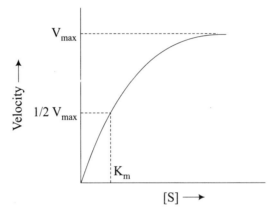

Fig. 7.3   Determination of K_m

The substrate concentration, at which half of the maximum velocity is obtained, is taken as K_m. However, this procedure is not always very satisfactory. Lineweaver – Burk proposed a simplified transformation of Michaelis-Menten equation into a linear equation:

$$\frac{1}{V} = \frac{K_m}{V_{max}} \cdot \frac{1}{[S]} + \frac{1}{V_{max}}$$

This gives a straight relationship between $\dfrac{1}{V_{max}}$ and $\dfrac{1}{[S]}$ when V is plotted against [S]. Such a plot is called a double-reciprocal plot or Lineweaver-Burk plot where reciprocal of V i.e. $\dfrac{1}{V}$ is plotted against reciprocal of [S] i.e. $\dfrac{1}{[S]}$ (Fig. 7.4).

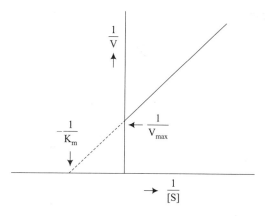

Fig. 7.4 Lineweaver-Burk plot

7.4 FACTORS AFFECTING RATE OF ENZYME ACTION

There are several factors which affect the rate of enzyme action. These factors are – temperature, pH, concentration of enzyme, concentration of substrates, activators and inhibitors.

(i) *Temperature:*   Each enzyme acts at a specific temperature which is called its *optimum temperature*. Most of the enzymes have an optimum temperature within the range of 35°-40°C. Generally, as the temperature of an enzymatic chemical reaction increases between 0°C to 45°C range, the activity of enzymes increases. It has been observed that for 10°C rise in temperature the rate of enzyme reaction increases by two times. This is known as Vant Hoff's law. If the temperature is further increased beyond 45°C, the activity of enzyme decreases. This is due to thermal denaturation of enzyme (Fig. 7.5).

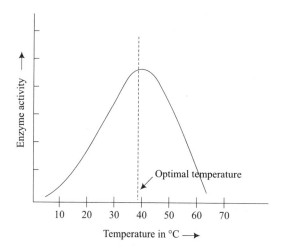

Fig. 7.5 Effect of temperature on enzyme activity

(ii) *pH:* The activity of the enzyme is also influenced by the pH of the medium. Each enzyme has an optimum pH at which its activity is maximum. Most of the enzymes are active between the pH range of 5 to 9 (Fig. 7.6). pH may alter the ionization of active sites or substrate which is required for ES complex formation.

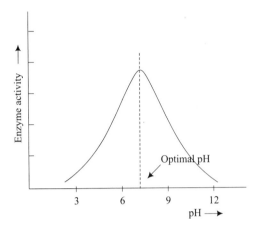

Fig. 7.6 Effect of pH on enzyme activity

(iii) *Concentration of enzyme:* The velocity of the enzymatic reaction is directly proportional to the concentration of enzyme, provided substrate is in large quantity (Fig. 7.7).

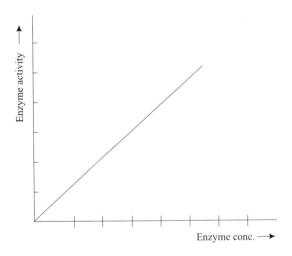

Fig. 7.7 Effect of concentration of enzyme activity

(iv) *Concentration of substrate:* Initially, the rate of enzyme action increases with the increase in substrate concentration. After reaching a certain point, there is no change in enzyme activity with the increase in substrate concentration. At this point V_{max}, the enzyme is saturated with the substrate and can function no faster (Fig. 7.8).

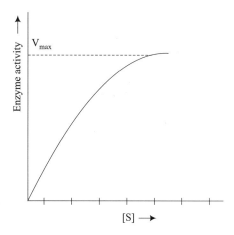

Fig. 7.8   Effect of concentration of substrate on enzyme activity

(v) *Activators:*   Activators are the metallic ions and coenzymes which activate the enzyme activity. The metals may help either in maintaining or producing active structural conformation of the enzyme. They may also help in the formation of ES complex.

(vi) *Inhibitors*:   Inhibitors are the substances which inhibit the enzyme action. This process is known as *enzyme inhibition*. Enzyme inhibition is of two types:
   (i) Competitive inhibition
   (ii) Non-competitive inhibition

7.4.1   Competitive Inhibition

This is a reversible type of inhibition in which inhibitor resembles the normal substrate in its three dimensional structure. The inhibitor competes with the substrate for binding to the active sites of the enzyme. In such reaction both ES complex and EI (Enzyme – inhibitor) complexes are formed. But EI complex does not form the products. Amounts of ES and EI complex depend upon the concentration of the substrate and inhibitor present and on the affinity between enzyme and substrate/inhibitor.

$$+I \diagup \text{EI complex} \longrightarrow \text{No reaction}$$
$$E$$
$$+S \diagdown \text{ES complex} \longrightarrow E + P$$

The rate of product formation depends on the concentration of ES complex. Thus, if inhibitor binds very tightly to the enzyme, there is little free enzyme available to combine with the substrate to form ES complex. If the affinity of the inhibitor towards enzyme is not strong enough, it will not decrease the rate of catalyzed reaction so markedly. Thus, in a reaction, if concentration of substrate is increased, it will increase the probability of the enzyme to combine with the substrate rather than with inhibitor and the rate of reaction will also rise. In competitive inhibition, the effect of inhibitor can be minimized by increasing the concentration of substrate. At a sufficiently high substrate concentration, the rate of reaction will be the same as in the absence of inhibitor. Inhibition of succinate dehydrogenase by malonate is an example of competitive inhibition. Succinate dehydrogenase catalyzes formation of fumarate from succinate. Malonate competes with succinate because both have similar configuration (Fig. 7.9).

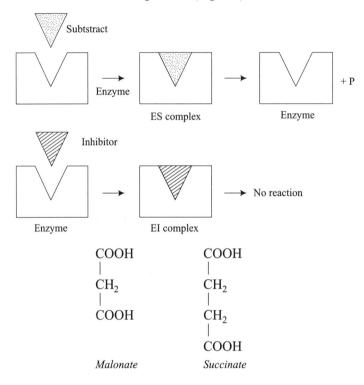

Fig. 7.9   Competitive inhibition

Many drugs work on the principle of competitive inhibition. Sulphonamides are very commonly used anti-bacterial agents. These are structurally similar to para-aminobenzoic acid (PABA) which is essential for synthesis of folic acid by enzyme action. Folic acid is needed for bacterial growth and survival. Sulphonamide drugs competitively inhibit enzyme action and folic acid is not synthesized and growth of bacteria is inhibited.

7.4.2   Non-competitive Inhibition

In non-competitive inhibition substrate and inhibitor do not compete for the active site of enzyme. Non-competitive inhibition is of two types:

(a) Reversible non-competitive inhibition

(b) Irreversible non-competitive inhibition

7.4.2.1 *Reversible inhibition*

In this case the inhibitor binds to a different region on the enzyme. The inhibitor binds either to free enzyme or to ES complex and ESI is formed. ESI may break down to form products at a slower rate than does ES and, thus reaction is slowed down (Fig. 7.10).

7.4.2.2 *Irreversible inhibition*

There are certain other substances which bind covalently with the enzyme or can destroy a functional group on the enzyme which is essential for enzyme activity. A variety of enzyme poisons-Ag^{2+}, Hg^{2+}, oxidising agents belong to this category. Since these inhibitors do not bear any structural resemblance to the substrate, an increase in substrate concentration does not affect the inhibition.

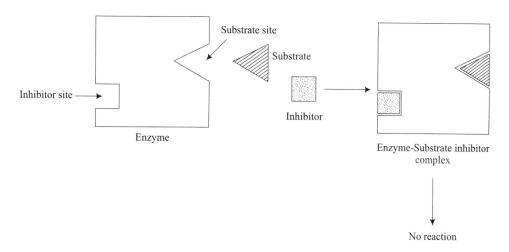

Fig. 7.10 Non-competitive inhibition (reversible)

7.5 ISOZYMES

Isozymes or isoenzymes are the multiple forms of an enzyme. They differ in amino acid sequence but catalyze the same chemical reactions. Isozymes were first described by *Hunter and Markert* (1957). They defined them as different variants of the same enzyme having identical functions and present in the same individual. These enzymes usually display different parameters (i.e. different K_m values).

They also differ in their chemical configuration, immunological tests and in their electro phoretic mobility. They are encoded by different gene loci which usually arise through gene duplication but can also arise from hybridization.

One of the important examples of isozymes is lactate dehydrogenase (LDH). LDH catalyzes the conversion of pyruvate and lactate. The enzyme occurs in five different forms in various tissues which differ in their electrophoretic mobility in starch gel. LDH consists of four polypeptide chains (tetramer). The polypeptide chains are of two types – M-chain and H-chain. M-chain is predominant in muscles while H-chain is predominant in heart tissues. Each subunit is the product of different gene. The five isozymes result from five possible combinations of these subunits:

HHHH   HHMM   MMMM

HHHM   HMMM

The relative portion of the five isozymes is the characteristic for each tissue and for each stage of embryonic development. The amino acid sequence is 75% identical. The H_4 enzyme found in heart, has a higher affinity for substrates than does M_4 enzyme. The high levels of pyruvate allosterically inhibit the H_4 but not the M_4 enzyme. The M_4 isoezyme functions optimally in an anaerobic condition whereas H_4 does so in aerobic environment.

Another example of isozymes is glucokinase, a varient of hexokinase. This has a different regulatory features and lower affinity for glucose as compared to other hexoses.

7.6 ZYMOGENS

Zymogens are the inactive precursors of the enzymes. A zymogen requires a biochemical change, like hydrolysis, to expose the active site to convert it into an active enzyme. The amino acid chain that is released upon activation is called the activation peptide. Many proteolytic enzymes of the stomach and pancreas are released in this form. Pepsin, trypsin and chymotrypsin are the examples of enzymes which are released in their inactive forms – pepsinogen, trypsinogen and chymotrypsinogen respectively. Specific cleavage causes conformational changes that exposes their active sites and convert them into their active forms.

Pepsinogen → Pepsin

(Inactive form)   (Active form)

Trypsinogen → Trypsin

(Inactive form)   (Active form)

Generally prefix 'pro' or suffix 'ogen' is added to an enzyme's name to indicate it as a precursor. Other examples include several proteins involved in blood coagulation and in the complement system.

For example, Fibrin, which is produced as fibrinogen, is converted into its active form by thrombin.

Fibrinogen $\xrightarrow{\text{Thrombin}}$ Fibrin

(Inactive form)   (Active form)

Thrombin itself is produced in its inactive form known as prothrombin.

Prothrombin → Thrombin

(Inactive form)   (Active form)

7.7 ALLOSTERIC ENZYMES

There are certain enzymes whose activity is influenced by certain other molecules. These are known as allosteric enzymes. The word 'allosteric' comes from the Greek words 'allos' – other and 'stereos' shape. Allosteric enzymes change their shape, or conformation upon binding of a modulator. An *allosteric modulator* (or effector) may be an allosteric inhibitor or allosteric activator. Often the modulator is a substrate itself. If the modulator and substrate are the same, the enzyme is called a *homotropic*. When the modulator is a molecule other than the substrate, the enzyme is said to be *heterotropic*.

The properties of allosteric enzymes are different from nonallosteric enzymes. They are generally larger and more complex than nonallosteric enzymes. Allosteric enzymes tend to have several subunits. In addition to active sites, allosteric enzymes have one or more regulatory or allosteric sites for binding the modulator. When the allosteric activator binds to the enzyme at the regulatory site, it makes certain changes in the shape of the enzyme so that it can bind its substrate more effectively and catalyzes the reaction.

In case of homotropic allosteric enzymes, where the substrate and modulator are the same, the same binding site, may function both as an active site as well as regulatory site. The binding of a substrate molecule enhances the binding of subsequent molecules to their active sites.

The allosteric effector may affect the enzymatic activity by two ways: It can either affect or (alter) V_{max} or K_m value. Enzymes whose K_m is altered by the effector are known as K-type enzymes and the enzymes whose V_{max} is altered by the allosteric effector are said to be V-type enzymes.

7.8 RIBOZYMES

It is well known that enzymes are protein in nature. But there are certain RNA molecules which possess enzymatic activity. These are known to catalyze some chemical reactions. These molecules are known as *ribozymes* (also called RNA enzymes or catalytic RNA). Many natural ribozymes catalyze either the hydrolysis of one of their own phosphodiester bonds, or the hydrolysis of bonds in other RNAs. Carl Woese, Francis Crick and Leslie Orgel were the first to suggest that RNA could act as a catalyst, in 1967. The first ribozymes were discovered by Thomas R. Ceck and Sidney Altman. They won the Nobel Prize in chemistry in 1989 for their discovery of catalytic properties of RNA.

Ribonuclease P (RNase P) is the most familiar ribozyme. It cleaves the head (5′) end of the precursors of tRNA molecules. In bacteria, RNase-P is known to have two components – a molecule of RNA and a polypeptide chain or protein called C5 protein. Both these components are required for the ribozyme to function properly *in vivo*. *In vitro*, RNA molecule can act alone as a catalyst. In eukaryotes, such as humans and yeast, RNase P consists of an RNA chain as well as nine to ten associated proteins. Recent studies have revealed that human nuclear RNase P is also required for the normal and efficient transcription of various small non coding RNA genes, such as tRNA, 5SrRNA, SRP RNA and U6 Sn RNA genes.

Some other naturally occuring ribozymes are – peptidyl transferase 23S rRNA, Group I and Group II introns, hair pin ribozyme, hammer head ribozyme, HDV ribozyme, mammalian CPEB 3 ribozyme, VS ribozyme, glm S ribozyme and Co TC ribozyme. Some artificially self cleaving RNAs have been produced which have good enzymatic activity.

Ribozymes may have a variety of applications. They provide a very useful means of studying gene function and gene discovery. They have an important role as therapeutic agents. Studies are going on to develop ribozymes as therapeutic agents against HIV and Cancer.

7.9 INDUSTRIAL USE OF ENZYMES

Enzymes have been used for the production of food products such as cheese, wine and vinegar and other commodities since ancient times. Today, a large number of enzymes are commercially used in the manufacturing of different useful products. They have many applications in different industries. More than 75% of industrial enzymes are hydrolases. They are used for the degradation of various natural substances. Proteases have extensive use in the detergent and dairy industries. Various carbohydrases are used in textile, detergent and baking industries. Some applications of enzymes in different industrial productions are as follows:

1. *Detergents:* The largest industrial application of enzymes is in detergents. Bacterial proteinases are the most important detergent enzymes. Amylases are used to remove starch based stains. New and improved engineered versions of the detergent enzymes have been developed which have enhanced activity at lower temperatures and alkaline pH. The enzymes have been developed by the combined use of microbial screening and protein engineering.

2. *Drink:* Enzymes have number of applications in drink industry. Enzymes are widely used in fruit juice manufacturing. Enzymes–pectinase, xylanase and cellulase are widely used in fruit juice industries for the liberation of juice from the pulp, Pectinases and amylases are used in juice clarification. The use of enzymes in wine production is well known.

3. *Textile:* Enzymes have many applications in textile industry. They are used at various steps in the manufacturing of cotton textiles. In the preparation of Denim, enzyme cellulose is widely used. The enzyme is used for dye-fading. Hydrogen peroxide and oxidative enzymes like laccase are used as bleaching agents to bleach textiles.

4. *Baking:* Enzymes are widely used in baking industries. Amylases are used in bread production. They are used to increase the softness of bread and to improve its quality. Enzyme proteinases and lipases are used for dough strengthening.

5. *Animal feed:* The use of enzymes as feed additives is well established. Xylanases and β-glucanases are used in wheat and barley based diets respectively. The addition of enzymes reduces viscosity and increases the absorption of the nutrients. The enzyme phytase, in animal feed helps to release phosphate from phytic acid present in plant based feed materials.

6. *Leather:* Enzymes are used in leather industry at different levels of leather manufacturing. Proteolytic and lipolytic enzymes are used to remove unwanted parts from the animal skin used for leather manufacturing. Enzymes are used in dehairing and dewooling phases which improves the quality of leather.

7. *Paper Industry:* Enzymes are used in paper industry in pulp bleaching. Enzyme xylanases are used in pulp bleaching. Enzyme cellulase is used to remove the ink. Enzymes are also used in modification of starch, used in paper making to improve the strength, stiffness and erasability of paper.

8. *Personal Care Products:* Enzymes are also being used in the production of personal care products. e.g. Proteinase and lipase are used for contact lens cleaning. Similarly hydrogen peroxide is used in disinfections of contact lenses. Enzymes are also used in tooth pastes, and skin and hair products.

Table 7.2 Various industrial applications of enzymes

Industry	Enzyme	Application
Detergent	Protease	Removal of protein stains
	Amylase	Removal of starch stains
	Lipase	Removal of lipid stains
Drink	Pectinase	Depectinization
	Xylanase	Mashing
	Amylase	Juice clarification
Textile	Cellulase	Denim finishing
	Peroxidase	Bleaching
	Amylase	De-sizing
Baking	Amylase	Bread softness
	Proteinase	Dough strengthening
	Lipase	Dough strengthening
Animal feed	Xylanase	Absorption of nutrients
	β-glucanase	Absorption of nutrients
	Phytase	Phosphate release
Leather	Protease	Dehairing and dewooling
	Lipase	Dehairing and dewooling
Paper Industry	Xylanase	Pulp bleaching
	Cellulase	Removal of ink.
Personal Care products	Proteinase	Contact lens cleaning
	Lipase	Contact lens cleaning
	Peroxidase	Antimicrobial.

7.10 SUMMARY

- Enzymes are the biocatalysts which are protein in nature.
- Enzymes may be classified into various groups on the basis of their substrate or the type of reaction they catalyze. IUB has classified enzymes into six classes.
- The mechanism of action of enzyme is explained by two models: Lock and key model and Induced fit model.
- Michaelis and Menten equation describes the relationship between initial reaction velocity and substrate concentration under steady condition.
- K_m is the concentration of substrate at which the rate of reaction is $1/2 V_{max}$
- The factors which affect the rate of enzyme action are temperature, pH, concentration of enzyme, concentration of substrate, presence of activators and inhibitors.

- Isoenzymes are the multiple forms of an enzyme.
- Some enzymes are released in inactive form. Such inactive form of an enzyme is known as zymogen or proenzyme.
- Certain RNA molecules are also known to possess enzymatic activity. Such RNA molecules are called ribozymes.
- The activity of allosteric enzyme is influenced by certain other molecules.
- Enzymes have a wide variety of applications in different industries.

EXERCISE

1. What are enzymes? Discuss their properties.
2. Explain the mechanism of enzyme action.
3. What is K_m value? Explain its significance.
4. Describe the industrial use of enzymes.
5. Write short notes on:
    (a) Ribozymes
    (b) Isozymes
6. Differentiate between:
    (a) Coenzyme and cofactor
    (b) Enzymes and catalysts
    (c) Competitive and non-competitive inhibition

Vitamins

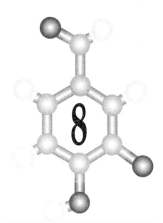

Vitamins are complex organic compounds that are required in small quantities for a variety of biochemical functions. These are generally not synthesized in the body and must, therefore, be supplied with food. Dr. Casimir Funk (1912) isolated from the rice polishings, an amine which was responsible for curing the deficiency symptoms of 'beri-beri'. As it was an amine and was responsible for the normal maintenance of health, he used the term '*Vitamine*' (vital amine). Today, the word is spelled as *Vitamin* since many of the substances of this class are not amines. Vitamins either participate in the production of coenzymes or act as regulators of biochemical processes.

8.1 CLASSIFICATION

Vitamins are classified into two large groups, on the basis of their solubility.

1. Water-soluble Vitamins, and
2. Fat-soluble Vitamins.

Vitamins B and C are water-soluble, whereas vitamins A, D, E and K are fat-soluble.

8.2 WATER-SOLUBLE VITAMINS

Water-soluble vitamins, because of their solubility in water, rarely accumulate in toxic concentrations in the body and excess of these are excreted in urine. Their storage is limited and they must be provided regularly.

Vitamins B and C are the water-soluble vitamins. Vitamins B is a group of several vitamins which include:

- Thiamine (vitamin B_1)
- Riboflavin (vitamin B_2)
- Niacin (vitamin B_3)
- Pantothenic acid (vitamin B_5)
- Pyridoxine (vitamin B_6)
- Biotin

- Cobalamin (vitamin B_{12})
- Folic Acid

8.2.1 Thiamin (B_1)

Thiamin was crystallized in 1925 by *B.C.P. Jansen* and its molecular structure was finally established by *R.R. Williams* and his colleagues in 1935.

Structure and functions

Thiamin consists of a substituted pyrimidine joined by a methylene bridge to a substituted thiazole. Thiamin is rapidly converted to its active form *thiamin pyrophosphate* (TPP) in the brain and liver by a specific enzyme, *thiamin diphosphotransferase*.

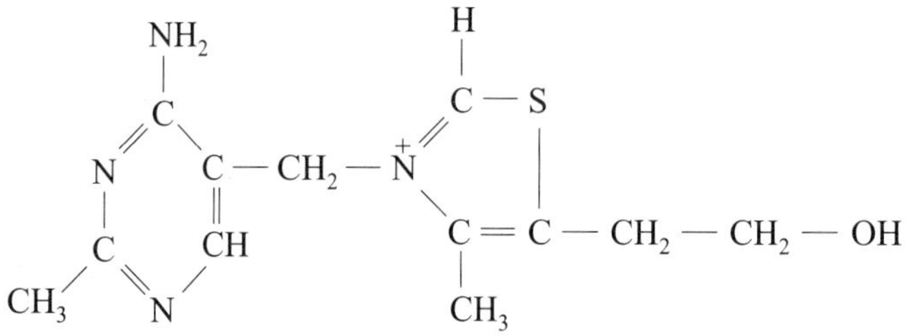

2, 5 Dimethyl
6-amino pyrimidine

4 methyl-5 hydroxy
ethylthiazole

Thiamin

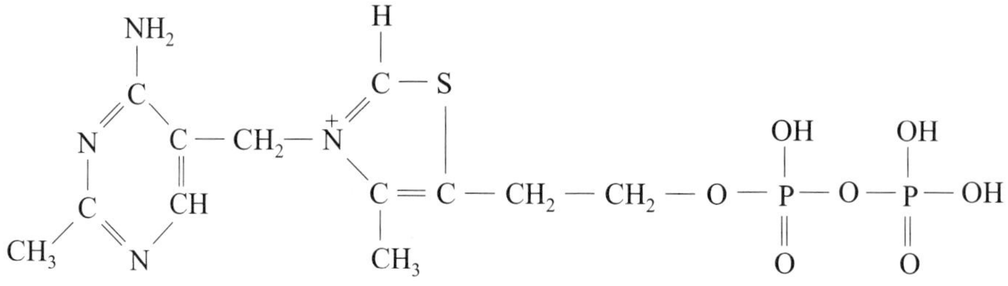

Pyrophosphate

Thiamin diphosphate (pyrophosphate)

Thiamin pyrophosphate serves as coenzyme for two classes of enzyme catalyzed reactions in the mainstream of carbohydrate metabolism in which aldehyde groups are removed and/or transferred:

(i) The decarboxylation of α-keto acids, and

(ii) The formation or degradation of α-ketols.

In these reactions, the thiazole ring of thiamin pyrophosphate serves as a transient carrier of a covalently bonded active aldehyde group. Mg^{2+} is also required as a cofactor.

Sources

Thiamin is present in almost all plants and animal tissues commonly used as food but the content is usually small. Unrefined cereal grains and meat are good sources of vitamin. The dietary requirement for thiamin is proportional to the caloric intake of the diet and ranges from 1.0-1.5 mg/day for normal adults.

Deficiency

The deficiency of thiamin causes *'beri-beri'*. It is caused by carbohydrate rich low thiamin diets such as polished rice or other highly refined foods. Beri - beri affects the peripheral nervous system, gastro-intestinal tract and the cardio-vascular system. Early symptoms include peripheral neuropathy, exhausation and anorexia which progress to edema and cardio-vascular, neurologic and muscular degeneration. There are two forms of disease: 'dry' and 'wet' beri-beri. In the *dry form* there is a rapid loss of weight and muscle wasting, marked peripheral neuritis and muscular weakness. Deep reflexes are lost and sensory changes may occur. In *wet beri-beri* there is a generalized edema that may lead to weakness and muscular wasting.

8.2.2 Riboflavin (B₂)

It is a coloured, fluorescent pigment that is relatively heat-stable but decomposes in the presence of visible light.

Structure and functions

The structure of riboflavin was established by *R. Kuhn* and *P. Karrer* in 1935. It consists of a heterocyclic isoalloxazine ring attached to the sugar alcohol, ribitol. The riboflavin acts in its two active forms:

- Flavin mononucleotide (FMN), and
- Flavin adenine dinucleotide (FAD)

FMN is formed by ATP- dependent phosphorylation of riboflavin where as FAD is synthesized by a further reaction with ATP in which the AMP moiety of ATP is transferred to FMN.

Flavin

Riboflavin

Riboflavin phosphate- FMN

Flavin adenine dinucleoide (FAD)

The flavin nucleotides function as prosthetic groups of oxidation-reduction enzymes known as *flavo enzymes* or *flavoproteins*. These enzymes function in the oxidative degradation of pyruvate, fatty acids and amino acids and also in the process of electron transport. Flavin nucleotides undergo reversible reduction of the isoalloxazine ring in the catalytic cycle of flavoproteins to yield the reduced nucleotides – $FMNH_2$ and $FADH_2$.

Sources

Riboflavin is synthesized by plants and micro-organisms but not by mammals. Yeast, liver and kidney are the good sources of the vitamin. The normal diet requirement for riboflavin is 1.2-1.7 mg/day for normal adult.

Deficiency

The deficiency of riboflavin does not lead to major life-threatening conditions. However, when there is deficiency of riboflavin it may lead to angular stomatitis, cheilosis, glossitis, seborrhea and photophobia. Riboflavin decomposes when exposed to visible light. This characteristic can lead to riboflavin deficiencies in newborns treated for hyperbilirubinemia by phototherapy.

8.2.3 Niacin (B$_3$)

Niacin is the name given for nicotinic acid and nicotinamide. Nicotinic acid is so named because it is a component of the toxic alkaloid nicotine of tobacco.

Structure and function

Nicotinic acid is a monocarboxylic acid derivative of pyridine. Niacin acts in the form of *nicotinamide adenine dinucleotide* (NAD), also called diphosphopyridine nucleotide (DPN), and *nicotinamide adenine dinucleotide phosphate* (NADP) also called as triphosphopyridine nucleotide (TPN). These two coenzymes are also referred to as *pyridine coenzymes* or *pyridine nucleotides.*

Nicotinic Acid	Nicotinamide

The pyridine nucleotides function as the coenzymes of a number of oxidoreductases. They act as electron acceptors during the enzymatic removal of hydrogen atoms from specific substrate molecules. One hydrogen atom from substrate is transferred as a hydride ion to the nicotinamide portion of the oxidized form of these coenzymes (NAD$^+$ and NADP$^+$) to yield the reduced coenzymes (NADH and NADPH). The other hydrogen from the substrate becomes a hydrogen ion.

Sources

Plants and most animals can make nicotinic acid from the amino acid tryptophan. Niacin is found most abundantly in yeast. Lean meats, liver, milk, tomatoes and several leafy vegetables are the good sources of niacin. The daily requirement for niacin is 13-19 niacin equivalents (NE) per day for a normal adult. One NE is equivalent to 1 mg of free niacin.

Deficiency

The deficiency of niacin causes *pellagra*. The symptoms of pellagra include weight loss, digestive disorders, dermatitis, depression and dementia. Several physiological conditions (e.g. Hartnup disease and malignant carcinoid syndrome) as well as certain drug therapies can lead to niacin deficiency.

8.2.4 Pantothenic Acid (B_5)

Panthothenic acid was first recognized by *R.J. Williams* as a growth factor for yeast. It is formed by plants and many bacteria but is required in the diet of vertebrates.

Structure and functions

Pantothenic acid is formed by combination of pantoic acid and β-alanine.

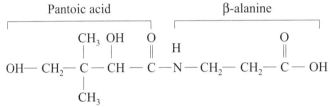

Pantothenic acid

The active form of pantothenic acid is coenzyme A which is essential to several fundamental reactions in metabolism and acyl carrier protein (ACP). ACP participates in reactions concerned with fatty acid synthesis.

Sources

Pantothenic acid is widely distributed in food, particularly in that of animal sources. Egg yolk, kidney, liver, yeast, whole grain cereals and skimmed milk are the good sources of this vitamin.

Deficiency

Deficiency of pantothenic acid, in humans, is not common due to its distribution in food. However, the deficiency may cause burning sensation, muscle weakness, abdominal disorders and general depression.

8.2.5 Pyridoxine (B_6)

Vitamin B_6 was first identified as essential in the nutrition of the rat for the prevention of a dermatitis called acrodynia. As it is antidermatic factor for rats, it is also known as *adermin*.

Structure and functions

Vitamin B_6 is a group of three closely related pyridine derivatives: *pyridroxine, pyridoxal* and *pyridoxamine* and their corresponding phosphates. Pyridoxine, pyridoxal phosphate and pyridoxamine phosphate are the main representatives of the vitamin in the diet and are interconvertible and have equal vitamin activity.

Pyridoxal phosphate is the active form of vitamin B_6. Pyridoxal phosphate is the major form transported in plasma.

It acts as the coenzyme of several enzymes of amino acid metabolism. The coenzyme is an integral part of the mechanism of action of phosphorylase, the enzyme mediating the breakdown of glycogen. Muscle phosphorylase may account for as much as 70-80% of total body vitamin B_6.

Pyridoxine   *Pyridoxal*   *Pyridoxamine*

Sources

The vitamin is widely distributed in nature. Liver, bananas, meat, vegetables and eggs are the good sources of vitamin B_6. The requirement for vitamin B_6 in the diet is proportional to the level of protein consumption ranging from 1.4-2.0 mg/day for a normal adult. During pregnancy and lactation the requirement for vitamin B_6 increases.

Deficiency

Deficiency of vitamin B_6 alone is rare. But deficiency may arise during lactation or as a result of malabsorption, alcoholism and drugs. Peculiar dermatitis (acrodynia), reduced growth, degeneration of the nerves, reproductive failure and hypochromic microcytic anaemia are the symptoms which occur as a result of deficiency of the vitamin.

8.2.6   Biotin

Kogl and *Tannis* isolated from egg-yolk a crystalline growth-factor for yeast in 1936. They named it as 'Biotin'.

Structure and functions

Biotin is an imidazole derivative. It functions as a component of specific multisubunit enzymes that catalyze carboxylase reactions.

Sources

Sufficient quantities of biotin are provided by intestinal bacterial synthesis to the mammals. Egg yolk, liver, kidney, yeast, peas, nuts are the good sources of the vitamin.

Biotin

Deficiency

Since a large portion of the human requirement for biotin is met by synthesis from intestinal bacteria, biotin deficiency is caused not by simple dietary deficiency but by defects in utilization. Consumption of raw eggs can cause biotin deficiency. The egg white contains a heat–labile protein, 'avidin' which combines very tightly with biotin and prevents its absorption and leads to biotin deficiency. Experimentally induced deficiency is manifested by mild dermatitis, anorexia and nausea. Children with this deficiency, sometimes, have immuno-deficiency diseases.

8.2.7   Cobalamin (B_{12})

Vitamin B_{12}, the first natural product containing cobalt, was first isolated by *Smith and Parker* in 1948 from liver.

Structure and functions

Vitamins B_{12} (Cobalamin) has a complex ring structure (corrin ring), similar to a porphyrin ring, to which is added a cobalt ion at its center. In the liver, the vitamin is found as methyl cobalamin, adenosylcobalamin and hydroxycobalamin. Methylcobalamin and deoxyadenosyl cobalamin are the active B_{12} coenzymes. The vitamin has a number of roles in animals. Vitamin B_{12} is absorbed from ileum. The absorption required it to be bound by a highly specific glycoprotein, 'intrisic factor' secreted by parietal cells of the gastric mucosa.

Sources

Vitamin B_{12} is synthesized exclusively by microorganimsm. Plants contain no vitamin B_{12} unless they are contaminated by microorganisms. Animal products like milk, liver, kidney, eggs and cheese are the primary dietary sources of this vitamin.

Deficiency

The deficiency of vitamin B_{12} leads to the development of *pernicious anemia* or macrocytic anemia. Pernicious anemia results when the absorption of vitamin B_{12} is prevented by lack of intrinsic factor. This vitamin is necessary for the formation and maturation of RBCs. Neurological complications are also associated with vitamin B_{12} deficiency and result from a progressive demyelination of nerve cells. The demyelination is thought to result from the increase in methylmalonyl-co A that results from vitamin B_{12} deficiency.

Cobalamin

8.2.8 Folic Acid

The name folic acid was suggested by *Mitchell*, *Snell* and *Williams* because of its isolation from spinach leaf. The chemical structure was established in 1945. 'Folacin' is the generic term for folic acid and related substances having the biochemical activity of folic acid.

Structure and functions

Folic acid consists of the base *pteridine* attached to one molecule each of p-aminobenzoic acid (PABA) and glutamic acid.

Folic acid derivatives in the diet are cleaved to monoglutamyl folate for absorption in the intestine. Most of this is reduced to tetrahydrofolate. Tetrahydrofolate is the active folate. It acts as a coenzyme in the transfer of formyl and hydroxy methyl groups in different biological systems. Folic acid also plays an important role in formation and maturation of RBCs.

Folic acid

Sources

Yeast, liver and leafy vegetables are the major sources of this vitamin.

Deficiency

The most pronounced effect of folate deficiency on cellular processes is upon DNA synthesis. The inability to synthesize DNA during erythrocytic maturation leads to abnormally large erythrocytes termed as macrocytic anemia. Folate deficiencies are rare due to the adequate presence of folate in the food. Poor dietary habits as those of chronic alcoholics can lead to folate deficiency. In non-alcoholics, impaired absorption or metabolism or an increased demand for the vitamin can lead to folate deficiency.

8.2.9 Vitamin C [Ascorbic Acid]

Vitamin C (Ascorbic acid), also known as *antiscorbutic* vitamin was isolated by *Szent Gyorgyi* (1928). It is a colourless crystalline substance, freely soluble in water. It is destroyed by heat, cooking etc.

Structure and functions

The structure of vitamin C resembles that of a monosaccharide. The structure of ascorbic acid is reminiscent of glucose, from which it is derived in the majority of mammals. Ascorbic acid can easily be oxidised to dehydroascorbic acid which itself can act as a source of the vitamin.

$$
\begin{array}{l}
\mathrm{O}\!=\!\mathrm{C} \!-\!\!\!-\!\!\!-\!\!\!-\!\!\!- \\
\quad | \qquad\qquad | \\
\quad \mathrm{C}\!-\!\mathrm{OH} \quad | \\
\quad \| \qquad\qquad \mathrm{O} \\
\quad \mathrm{C}\!-\!\mathrm{OH} \quad | \\
\quad | \qquad\qquad | \\
\mathrm{H}\!-\!\mathrm{C} \!-\!\!\!-\!\!\!-\!\!\!-\!\!\!- \\
\quad | \\
\mathrm{OH}\!-\!\mathrm{C}\!-\!\mathrm{H} \\
\quad | \\
\quad \mathrm{CH_2OH}
\end{array}
$$

Ascorbic acid

The main function of ascorbic acid is as a reducing agent in a number of reactions. Vitamin C has the potential to reduce cytochromes 'a' and 'c' of the respiratory chain as well as molecular oxygen.

In collagen synthesis, ascorbic acid is required for hydroxylation of proline. It is, therefore, required for the maintenance of normal connective tissue as well as wound healing. Ascorbic acid is also necessary for bone remodelling due to the presence of collagen in the organic matix of bones. Vitamin C is also believed to be involved in steroidogenesis since adrenal cortex contains large amount of vitamin C which is rapidly depleted when the gland is stimulated by adrenocortico trophic hormones. Absorption of iron is significantly enhanced in the presence of vitamin C.

Ascorbic acid may act as a general water soluble anti-oxidant and may inhibit the formation of nitrosamines during digestion.

Sources

Citrus fruits (lime, lemon, orange) are the best food sources of vitamin C. Animal sources are generally poor except the adrenal cortex.

Deficiency

The deficiency of vitamin C causes the disease 'scurvy.' This is due to the role of vitamin C in the post-translational modification of collagens. Scurvy is characterized by easily bruised skin, muscle fatigue, soft swollen gums, decreased wound healing and haemorrhaging, osteoporasis and anemia. Vitamin C is readily absorbed and so the primary cause of vitamin C deficiency is poor diet and/or increased requirement.

8.3 FAT-SOLUBLE VITAMINS

8.3.1 Vitamin A

Vitamin A was discovered by *Mc Collum* and *Davis*. Its chemical structure was determined by *Karrer* in 1931.

Structure and functions

Vitamin A or retinol is a polyisoprenoid compound containing a cyclohexenyl ring.

Retinol (Vitamin A)

In vegetables, vitamin A exists as a provitamin in the form of a yellow pigment β-*carotene*. β-carotene consists of two molecules of retinol joined at the aldehyde end of their carbon chain. In the body the main functions of Vitamin A are carried out by retinol and its derivatives *retinal* and *retinoic acid*. Retinol and retinal are interconverted in the presence of NAD or NAD requiring dehydrogenases or reductases.

The main function of vitamin A is its role in vision. Retinal is a component of the visual pigment *rhodopsin* which occurs in the rods of retina of eye. These are responsible for vision in poor light. Rhodopsin is made up of a protein 'opsin' and tightly bound 11-cis- retinal. When rhodopsin receives light, the 'cis-retinal' is converted to 'trans-retinal' by a photochemical non-enzymatic process. This causes the dissociation of opsin part from retinal. This dissocation induces nerve impulses which are the basis of vision.

The maintenance of the integrity of epithelial tissue is an important function of vitamin A. Vitamin A combines with certain structural proteins in the cells to stablize them.

Vitamin A is also required for the normal fertility. In experimental rats, vitamin A deficient male rats do not develop testes properly and sperm maturation does not take place. Retinoic acid is found to be involved in glycoprotein synthesis and this accounts for the action of vitamin A in promoting growth and differentiation of tissues.

Retinoids and cartenoids, both are known to have anticancer activity. Experiments have shown that retinoid administration diminishes the effect of certain carcinogens.

Sources

Fish liver oil, liver, egg-yolk, butter are good sources of vitamin A. All yellow pigmented vegetables and fruits–carrot, pumpkin, papayas, tomatoes etc. supply provitamin-carotene in the diet. Adult male and female require about 3000 I.U. of vitamin A per day. One I.U. is equivalent to 0.3 μg of retinol.

Deficiency

Deficiency of vitamin A leads to night blindness. Further depletion leads to karatinization of epithelial tissues of the eye, lungs, gastrointestinal and genitourinary tract coupled with reduction in mucous. Deficiency of this vitamin results in slowing of endochondrdrial bone formation and decreased osteoplastic activity.

8.3.2 Vitamin D

Vitamin D (Calciferol) is also known as antirachitic vitamin since the deficiency of this causes rickets, in children. The vitamin was discovered by *Mc Collum* in 1922. The main two forms of the vitamin are D_2 (*ergocalciferol*) and D_3 (*cholecalciferol*).

Structure and functions

Vitamin D is generated from dehydrocholesterol by the action of sunlight. Dehydrocholesterol occurs in animals and when the skin is exposed to sunlight, dehydrocholestrol is converted into cholecaliciferol. Similarly, in plants ergosterol is converted into ergocalciferol by U.V. radiations.

7-dehydrocholesterol
(Animals)

Cholecalciferol
(Vitamin D$_3$)

Vitamin D is involved in metabolism of calcium and phosphorus. It facilitates the absorption of phosphates. It also facilitates the transport and absorption of calcium from intestine. This is mediated through protein synthesis, since in experimental cases addition of actinomycin D (an inhibitor of mRNA Synthesis) inhibits vitamin D action. It is also involved in cell differentiation immune function.

Ergosterol
(Plants)

Ergocalciferol (vitamin D_2)

Sources

Fish liver oil is the best source of vitamin D. Butter, milk and egg also contain vitamin D. For an adult man daily requirement of vitamin D is about 2.5 μg.

Deficiency

The deficiency of vitamin D causes rickets in growing children and osteomalacia in adults. These occur due to softening of bones because of lack of calcium and phosphorus.

8.3.3 Vitamin E

Vitamin E or tocopherol was isolated by *Evans* and *Emerson* in 1936. The word tocopherol is derived from the Greek words, *tokos* means child birth and *phero* means to bear. Tocopherols are the antisterility factors in rats.

Structure and functions

Several types of tocopherols are active as vitamin E. They are methyl derivatives of the parent compound *tocol*. They differ from each other in the position of methyl groups. The four main types of tocopherols are:

1. α-tocopherol – 5,7,8 trimethyl tocol.
2. β-tocopherol – 5,8 dimethyl tocol.
3. γ-tocopherol – 7,8 dimethyl tocol.
4. δ-tocopherol – 8 methyl tocol.

Out of these, α-tocopherol has the highest natural distribution and biological activity.

The most striking chemical characteristic of vitamin E is its antioxidant property. Vitamin E appears to be the first line of defence against peroxidation of polyunsaturated fatty acids. The antioxidant action of tocopherol is effective at high oxygen concentration. In rats, vitamin E is required for the normal reproductive functions. In male rats the deficiency of vitamin E may lead to the destruction of germinal epithelium of testes. In female rats, the ovary is unaffected by vitamin E deficiency, but the foetus does not develop normally.

α-tocopherol

Sources

Wheat gram, sunflower seed, cotton seed, corn and soyabean oil are the rich sources of vitamin E. The daily requirement of the vitamin in adult is 20-25 I U.

Deficiency

Deficiency of vitamin E may give rise to anemia of the new born. The deficiency of this vitamin may cause muscular dystrophy and peripheral vascular disorders.

8.3.4 Vitamin K

Vitamin K was identified by *Dam* in 1935 as a factor, present in green leaves, which prevented haemorrhage. This is also known as antihaemorrhagic vitamin or coagulation vitamin.

Structure and functions

Vitamins belonging to K-group are polyisoprenoid substituted napthoquinones. There are two major forms of vitamin K. *Phylloquinone* (K_1) is the major form of vitamin K found in plants while *Menaquinone* (K_2) is found in animal tissues and is synthesized by bacteria in the intestine.

Vitamin K has been shown to be involved in the maintenance of normal levels of blood clotting factors II, VII, IX and X, all of which are synthesized in the liver. This vitamin has also some role in the synthesis of bone protein-osteocalcin. Vitamin K is a necessary cofactor in oxidative phosphorylation being associated with mitochondrial lipids. UV irradiation of mitochondria destroys their vitamin K content and ultimately their ability for oxidative phosphorylation.

Sources

Alfalfa, cabbage, cauliflower, spinach are the good sources of vitamin K_1. Vitamin K_2 is the metabolic product of normal intestinal bacteria of most of the higher animals.

Table 8.1 Summary of various vitamins, their sources and their deficiency diseases

Vitamin	Solubility	Common Sources	Deficiency diseases/ symptoms
A, Retinol	Fat-Soluble	Yellow pigment Vegetables and fruits - carrot, pumpkin, papayas, carrot, liver, egg-yolk	Night blindness, Xerophthalmia, bone abnormalities.
D, Calciferol	Fat-Soluble	Fish-liver oil, milk, egg	Rickets, Osteomalacia
E, Tocopherol	Fat-Soluble	Wheat gram, sunflower, cotton seed, soyabean oil and animal tissues	Sterility, anemia, muscle disorders
K, Antihaemorrhagic	Fat-Soluble	Alfalfa, cabbage, spinach	Delayed blood clotting
B_1 Thiamin	Water Soluble	Cereal grains, meat, egg	Beri-beri
B_2 Riboflavin	Water Soluble	Yeast, liver, egg	Dermatitis
B_3 Niacin	Water-Soluble	Milk, tomato, liver	Pellagra
B_5 Pantothenic acid	Water-soluble	Egg yolk, liver, yeast, whole grain, skimmed milk	Burning sensation, muscle weakness
B_6 Pyridoxine	Water- Soluble	Egg, yolk, liver, yeast, cereal, grains	Neurological disorders, dermatitis
Biotin	Water- Soluble	Liver, kidney, milk, egg yolk, legumes and grains	Very rare, with avidin, dermatitis
Cobalamin	Water- Soluble	Liver, egg, fish meat, kidney, milk	Pernicious anaemia
Folic acid	Water-Soluble	Yeast, liver, kidney, green leafy vegetables	Anaemia
C, Ascorbic acid	Water-Soluble	Citrus, fruits-orange, lemon, cabbage, tomatoes	Scurvy

Deficiency

Deficiency of vitamin K is not common, since most common foods contain this vitamin. In addition, intestinal flora of microorganisms synthesize this vitamin. However, the deficiency may occur due to malabsorption of fat or prolonged use of antibiotics or sulfa drugs. The deficiency of this vitamin causes profuse bleeding from minor wound and disturbances in blood clotting.

Phylloquinone (Vitamin K$_1$)

Menaquinone or faruoquinone (Vitamin K$_2$)

8.4 SUMMARY

- Vitamins are the complex organic compounds, required in small amounts for various biochemical reactions.
- On the basis of their solubility, vitamins are classified into two groups— water soluble and fat-soluble vitamins.
- Vitamin B and C are water-soluble and vitamins A, D, E and K are fat soluble.
- Vitamin B is a group of several vitamins. They are required for various metabolic activities and may also act as coenzymes.
- Vitamin C may act as a general water soluble anti-oxidant.
- Vitamin A is required for normal vision.
- Vitamin D is synthesized by animals on exposure of skin to sun light.
- Vitamin E has its role in reproduction.
- Vitamin K is the antihaemorrhagic vitamin.

EXERCISE

1. What are B-complex vitamins? Describe their physiological functions.
2. Describe the chemistry, functions and deficiency manifestations of vitamin C.
3. Name the fat soluble vitamins. Describe the functions of vitamin D.
4. Write the sources, functions and deficiency manifestations of vitamin A.
5. Write short notes on:
   (a) Thiamine
   (b) Riboflavin
   (c) Ascorbic acid
   (d) Calciferol
   (e) Antihaemorrhagic vitamin

Hormones

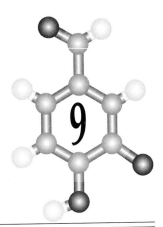

9

Hormones are the chemical substances secreted by endocrine glands into the blood and are carried to distant target organs and tissues to modify their structures and functions. The word 'hormone' is derived from Greek word '*hormaein*' meaning 'to excite'. Normally, hormones are stimulating substances which act as body catalysts. But all hormones are not excitatory. For example, somatostatin is a hormone which inhibits the secretion of a number of other hormones. Hormones catalyze and control diverse metabolic processes. Besides the endocrine glands, several other glandular tissues are considered to secrete hormones. For example, hormone *erythropoietin* is produced by kidney and regulates erythrocyte maturation. Some hormones are also produced by certain specialized cells of gastrointestinal tract. A summary of the hormones and their functions is given in Table 9.1.

Table 9.1 Common hormones and their functions

Source	Hormone	Functions
1. Hypothalamus	• Thyrotropin Releasing factor (TRF)	• Release of TSH from pituitary
	• Corticotropin Releasing factor (CRF)	• Release of ACTH from pituitary
	• Somatostatin	• Release (inhibits) of GH from pituitary
	• Gonadotropin Releasing factor (GRF-FSHRF, LHRF)	• Release of FSH and LH from pituitary
2. Pituitary Gland	• Thyroid stimulating hormone (TSH)	• Stimulates thyroid to secrete thyroxine.
a. Adenohypophysis	• Adrenocorticotrophin hormone (ACTH)	• Stimulates adrenal cortex to secrete corticoids.
	• Growth hormone (GH)	• Growth and metabolism
	• Follicle stimulating hormone (FSH)	• Maturation of follicles in females
		• Stimulates spermatogenesis in males.
	• Luteinizing hormone (LH)	• Luteinization, secretion of sex hormones.
	• Prolactin	• Milk secretion.

Contd.

b. Intermediate lobe	• Melanocyte stimulating hormone (MSH)	• Dispersion of melanophores.
c. Neurohypophysis	• Oxytocin	• Contraction of uterine muscles • Milk secretion.
	• Vasopressin (Anti diuretic hormone ADH)	• Renal reabsorption of water
3. Thyroid Gland	• Thyroxine	• Increases cell metabolism. • Regulates growth and development.
	• Calcitonin	• Regulation of Ca level in blood.
4. Parathyroid	• Parathormone	• Maintenance of Ca and P level.
5. Adrenal Gland	• Glucocorticoids	• Carbohydrate metabolism.
	• Mineralocorticoids	• Metabolism of minerals • Salt retention.
	• Adrenaline	• Break down of glycogen (stress hormone)
	• Nor-adrenaline	• Emergency hormone.
6. Pancreas	• Insulin • Glucagon	• Regulation of blood sugar level
7. Testis	• Testosterone	• Normal development of male reproductive organs, secondary sex characters.
8. Ovary	• Estrogen	• Normal development and functions of female reproductive organs, secondary sex characters.
	• Progesterone	• Regulation of menstruation • Maintenance of pregnancy.
9. Placenta	• Human chorionic gonadotrophic hormone (hCG)	• Maintenance of pregnancy.
	• Chorionic somatomammotropin (placental lactogen)	• Like that of GH and prolactin.
10. Kidney	• Erythropoietin	• Maturation of erythrocytes.
11. Gastro in estinal tract	• Gastrin	• Secretion of HCl
	• Pancreozymin	• Secretion of pancreatic juice
	• Cholecystokinin	• Stimulates pancreatic enzyme and HCO_3^- secretion.
	• Glucagon like peptide I	• Stimulates insulin release, inhibits glucagon release.

9.1 PROPERTIES OF HORMONES

Hormones are the chemical substances produced by endocrine glands into the blood stream. They are carried by the blood to their target organs and tissues. They are produced in minute amount and act as catalysts. A single hormone may have multiple effects on a single or on several target tissues.

Through hormones act as body catalysts resembling enzymes in some aspects but they differ from enzymes as they are produced in an organ other than that in which they ultimately perform their action. Structurally, like enzymes, they are not always proteins. Hormones act very specifically on certain organs and show a high degree of target specificity. One of the basic characteristics of hormones is the *feed back regulation* of their secretion. The complexity of these feed back regulatory systems and the mechanism involved vary greatly. But they tend to maintain homeostatic balance with regard to composition of body fluids, rates of various metabolic processes and biological functions.

9.2 CLASSIFICATION OF HORMONES

Chemically hormones can be classified into two major groups:

1. *Steroid hormones*: These are steroid in nature such as sex hormones (Testosterone, Estrogen, Progesterone) and hormones secreted by adrenal cortex.
2. *Amino acid derivatives*: These may be further divided into following categories:
   (i) Glycoproteins – LH, TSH
   (ii) Proteins – GH
   (iii) Polypeptides – ACTH, Insulin, Glucagon
   (iv) Modified amino acids – epinephrine (Adrenaline), nor-adrenaline.

9.3 MECHANISM OF ACTION OF HORMONES

Hormones are produced in the blood and are circulated throughout the whole body. They act specifically at their target organs/tissues. They control and catalyze diverse metabolic processes. Steroid and protein hormones differ from each other in their mechanism of action at their target cells. Peptide hormones have a particular mode of action by their interaction with specific macromolecular receptors present on the plasma membrane at the cell surface. By contrast, the steroid hormones pass through the plasma membrane and initiate their actions by binding to cytoplasmic receptors.

9.3.1 Action of Steroid Hormones

Steroid hormones act mostly by changing the transcription rate of specific genes in the nuclear DNA. They have been shown to bind to specific receptor molecules in cytoplasmic fraction of cells before translocation to nuclei and to initiate nuclei mediated responses. Steroid hormone receptors are specific oligomeric protein molecules in the cytoplasm of the cell. The binding of the hormones with their specific receptors results in a change in the conformation of the receptor molecule. The receptor is activated and is capable of binding to nuclear sites. The receptor-steroid complex is translocated to the nuclear chromatin and binds to a steroid-recognizing receptor site on the target cell genome. These sites-'acceptor sites' on target genome are believed to be constituted of nuclear acidic proteins and back bone of DNA molecule. The interaction of receptor-steroid complex with nuclear acceptor site initiates a sequence of nuclear related events. The consequent change in the

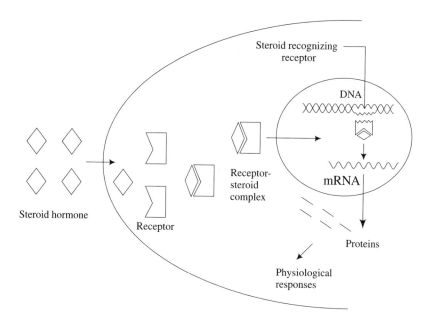

Fig. 9.1   Mechanism of action of steroid hormone

intracellular concentration of mRNA alters the rate of synthesis of a structural, enzymatic, carrier or receptor protein. This results in ultimate cellular effects. The receptor-steroid complex leaves the acceptor site as the free receptor and the steroid (Fig. 9.1).

9.3.2   Action of Protein Hormones

The receptors for the protein hormones are present on the plasma membranes of the cells. The actions of these hormones are believed to be mediated by the second messenger molecule, cyclic adenosine monophosphate (cAMP). The binding of the peptide hormones to their receptors on the cell surface activates membrane bound enzyme adenylate cyclase. Adenylate cyclase catalyzes the conversion of ATP to cAMP. cAMP, synthesized from ATP, interacts with at least two types of enzymes in the cell. It can be rapidly degraded by enzyme phosphodiesterase to 5' AMP and is devoid of any activity as second messenger or it can bind protein kinase to proceed with its normal action. The enzyme protein kinase is made up of two subunits i.e. a regulatory subunit and a catalytic subunit. cAMP has been shown to bind the regulatory subunit of this enzyme which results in the activation of catalytic subunit of the enzyme. Activation of the enzyme protein kinase is the major mechanism by which cAMP functions as a secondary messenger. Protein kinase phosphorylates a variety of proteins in different tissues. The phosphorylated proteins are presumed to be activated or deactivated to mediate various cellular functions (Fig. 9.2).

Just like cAMP, some other compounds like 1, 4, 5 inositol triphosphate (ITP) and diacyl glycerol also act as second messenger in some hormones like vasopressin, TRH etc. In case of these hormones, binding of the hormones to their surface receptors activates an enzyme phospholipase C, present on the inner surface of the membrane. This enzyme hydrolyses phoshatidyl inositol

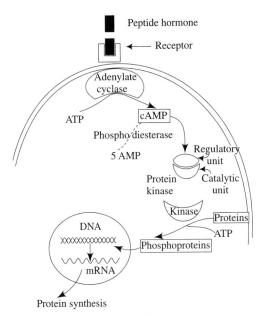

Fig. 9.2 Mechanism of action of protein hormones

phosphate to form 1, 4, 5 inositol triphosphate and diacylglycerol, which act as second messengers. Inositol triphosphate increases the level of Ca^{2+} in the cytosol pool by enhancing the mobilization of Ca^{2+} from mitochondria and endoplasmic reticulum. The calcium evokes a variety of responses and acts as another level of messenger. It is involved in the regulation of several enzymes which have special biochemical roles. Many of its effects are mediated through the proteins like calmodulin and troponin. Diacylglycerol activates the enzyme protein kinase C. This enzyme, then, phosphorylates specific enzymes and other proteins in the cytosl to mediate their activities.

9.4   FEED BACK REGULATION OF HORMONES

The concentrations of hormones in the body remain under control. Whenever there is increase or decrease in the concentrations of hormones, hormones themselves control their level. The control is through feed-back mechanism. Whenever there is high blood level of a target gland hormone, the hormone tends to suppress its further release. The hormone, or one of its products has negative feed back effect to prevent the secretion of the hormone or over activity at the target tissue. For example, thyroid gland secretes thyroxine under the influence of anterior pituitary hormone – TSH. TSH, in turn, is controlled by hypothalamic hormone TRF. Whenever, the levels of thyroxine increase beyond a limit, it has a negative effect at the level of hypothalamus or pituitary or both (long-loop feed back). The effect may also be at the level of thyroid itself (short-loop feed back) (Fig. 9.3). The feed-back regulation may be positive, in case, the level of hormone decreases beyond a certain limit.

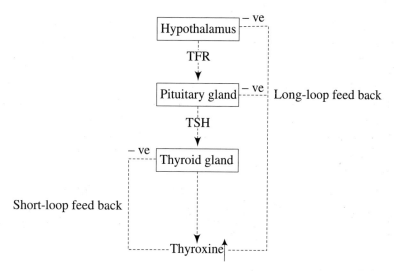

Fig. 9.3   Feed back mechanism

9.5   SUMMARY

- Hormones are the chemical secretions of endocrine glands.
- Hormones play a variety of roles in the organisms.
- Chemically, hormones are steroid or protein in nature.
- Hormones are very specific in nature. They act on specific target cells through their receptors. Their receptors are present either on the cell surface or inside the cell in the cytoplasm, depending on the nature of the hormone.

EXERCISE

1. What are hormones? Describe their properties.
2. Explain the mechanism of action of steroid hormones.
3. Write short notes on:
     (i) Pituitary hormones
     (ii) Feed-back mechanism.

Alkaloids

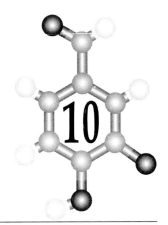

Alkaloids are the basic nitrogenous plant products containing nitrogen heterocycles in their structures. The term alkaloid, meaning alkali like, was first proposed by C.F.W. Meissner in 1819. Today more than 4500 alkaloids are known to occur in more than 4000 species of plants.

Alkaloids are mostly found in vascular plants. They rarely occur in Cryptogams (exception-ergot alkaloid), gymnosperms or monocotyledons. They occur abundantly in the families like apocynaceae, papaveraceae, papilionaceae, rubiaceae, rutaceae, ranunculaceae and solanaceae. Alkaloids have been isolated from the roots, leaves, bark of the stem or seeds of the plants. As they are basic in nature, in plants they exist as salts of basic acids such as acetic acid, malic acid, lactic acid and tartaric acid, in the free state, or as N-oxides.

The exact function of alkaloids in plants is not known. But they cause marked physiological actions when administered to animals. They function as strong poisons, anesthetics, pain relievers, narcotics etc. when treated in man and animals.

10.1 GENERAL PROPERTIES

The alkaloids are usually colourless, crystalline and non-volatile compounds. They are generally white solids (exception-nicotine is a brown liquid). They are soluble in organic solvents but are insoluble in water. However, some alkaloids are soluble in water. They are bitter in taste. Most of the alkaloids are optically active and are levorotatory. Alkaloids form salts with inorganic as well as organic acids. The salts formed by these alkaloids are soluble in water but insoluble in organic solvents. They give a precipitate with heavy metal iodides. Most of the alkaloids are precipitated from neutral or slightly acidic solution by Mayer's reagent (potassiomercuric iodide solution). Dragendorff's reagent (solution of potassium bismuth iodide) gives orange coloured precipitate with alkaloids.

10.2 CLASSIFICATION OF ALKALOIDS

Although there is no specific classification for alkaloids, but alkaloids can be classified according to *the plant genera in which they occur* or on the basis of *their chemical structure*.

On the basis of the chemical nucleus present in their structure, the following are the important groups of alkaloids:

- **Phenylethylamine alkaliod**-Ephedrine
- **Pyrollidine alkaloids**-Hygrine
- **Pyridine or piperidine alkaloids**-Ricinine, Coniine, Piperine
- **Pyridine-pyrrolidine alkaloids**-Nicotine
- **Tropane alkaloids**-Atropine, Cocaine
- **Quinoline alkaloids**-Quinine, Cinchonine
- **Isoquinoline alkaloids**-Papaverine, Narcotine, Berberine
- **Phenanthrene alkaloids**-Morphine
- **Indole alkaloids**- Ergotamine, Reserpine
- **Tropolone alkaloids**-Colchicine

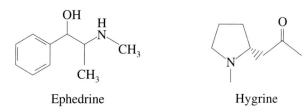

Ephedrine

Phenylethylamine alkaloid

Hygrine

Pyrollidine alkaloids

Piperine

Coniine

Pyridine or piperidine alkaloids

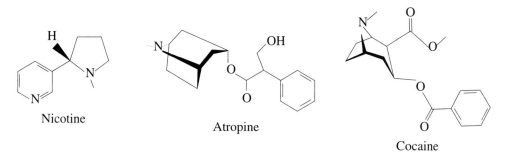

Nicotine

Atropine

Cocaine

Pyridine-pyrrolidine alkaloids

Tropane alkaloids

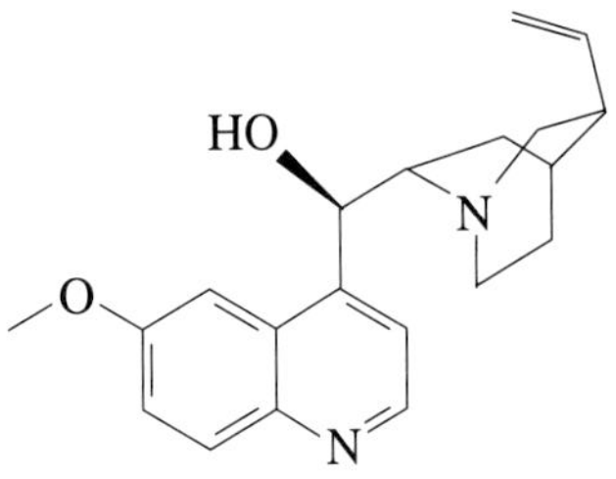

Quinine

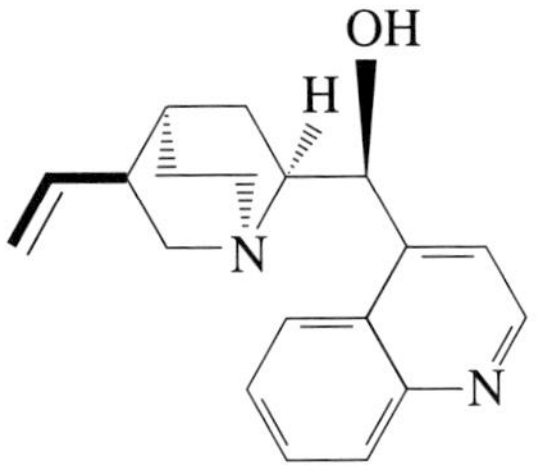

Cinchonine

Quinoline alkaloids

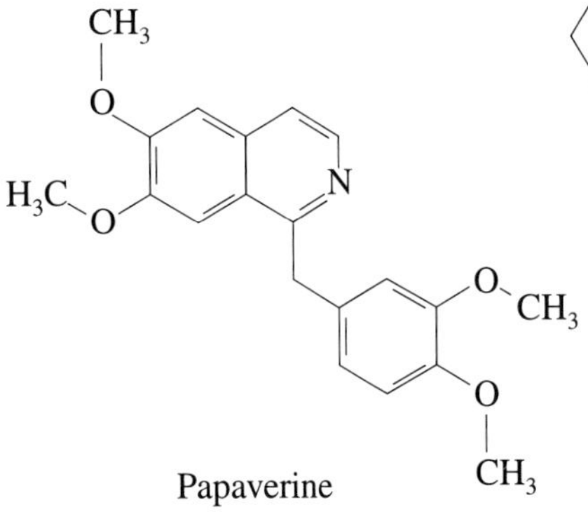

Papaverine

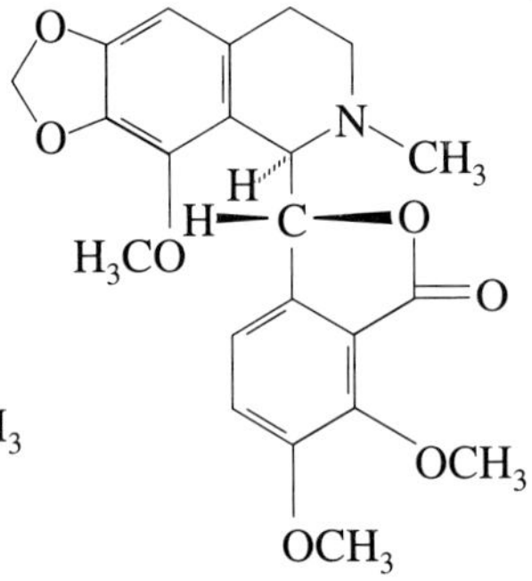

Narcotine

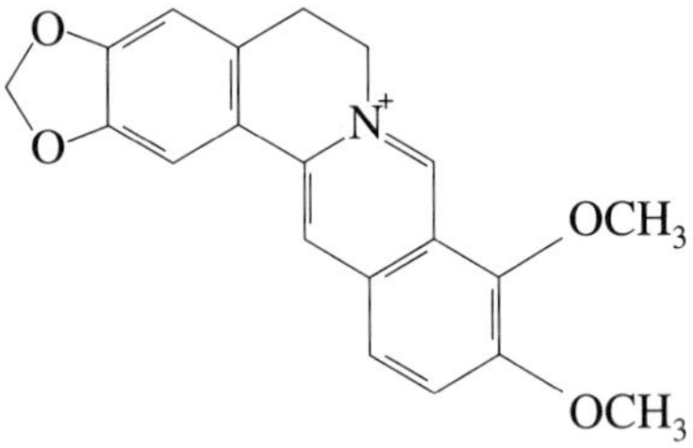

Berberine

Isoquinoline alkaloids

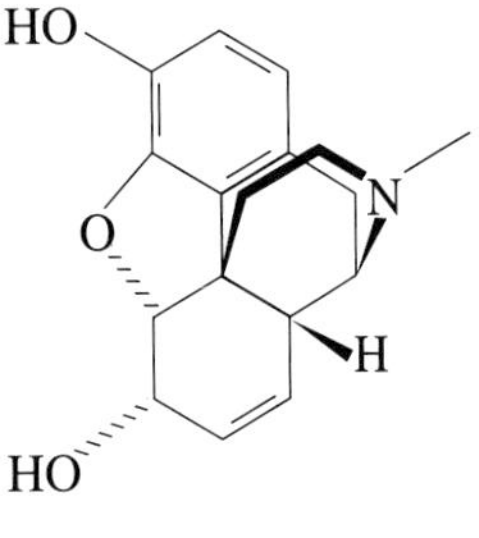

Morphine

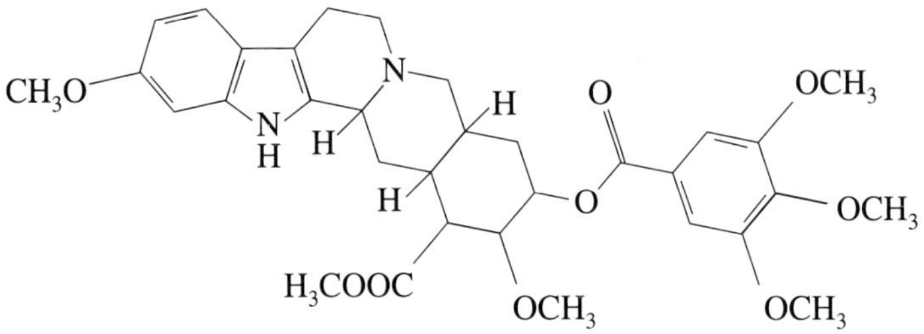

Reserpine

Phenanthrene alkaloids

Indole alkaloids

Colchicine

Tropolone alkaloids

10.3 BIOLOGICAL SIGNIFICANCE

Alkaloids have various physiological effects in men and other animals when they are treated with them.

In plants various functions have been attributed to the alkaloids. Alkaloids act as poisonous substances which afford plants safety from herbivores and insects. For example, in barley, the alkaloid *gramine* is an important factor in resistance to aphid *Schizaphis* sp. It has been observed that with increasing in concentration of nitrates in the nutrient solution, leaf *gramine* content increases and at the same time the infestation by the aphid *Schizaphis gramine* decreases.

It is also suggested that they are excretory products of the plants and excess of ammonia is excreted by this way. It is also believed that alkaloids may work as a nitrogen reserve in plants.

Alkaloids have chelating properties and is in this way they behave like growth regulators. There is some evidence that morphine plays role in the formation of viable seeds in poppy. Some alkaloids also inhibit enzyme activity.

10.4 SOME COMMON ALKALOIDS

10.4.1 Coniine

Coniine is a simple alkaloid which occurs in the seeds and other parts of the spotted hemlock *Conium maculatum*. The molecular formula of coniine is $C_8H_{17}N$. It is made up of pyridine ring with a side chain in the second position. It is a colourless, alkaline liquid and is highly poisonous. It is sparingly soluble in water but readily soluble in alcohol. It has an unpleasant odour and a burning taste. Both coniine and its salts are exceedingly poisonous causing death by paralyzing the nervous and respiratory system.

Coniine

10.4.2 Piperine

Piperine occurs in black pepper *Piper nigrum*. It is responsible for the sharp taste of the pepper. It is a colourless crystal but is

Piperine

much less toxic. The molecular formula of piperine is $C_{17}H_{19}NO_3$. It is formed by uniting piperic acid with piperidine through an amide group. It is sparingly soluble in water and forms salts with strong acids. Piperine is used as a flavouring additive in brandy and as an insecticide for houseflies.

10.4.3 Nicotine

Nicotine is the principal alkaloid of tabacco, *Nicotina tobacum* which occurs in the dry leaves. It is a colourless oily liquid. The molecular formula of the nicotine is $C_{10}H_{14}N_2$. It contains a pyridine nucleus having a side chain called N-methyl pyrrolidine. It is soluble in water and organic solvents. Nicotine is deadly poisonous. 30 to 50 mg nicotine when taken orally kills a man within a few seconds owing to paralysis of the nervous system including the respiratory centers. It is used as an insecticide and fungicide.

Nicotine

10.4.4 Quinine

It is commonly known as cinchona alkaloid. It is obtained from the bark of *Cinchona officinalis*. The molecular formula of quinine is $C_{20}H_{24}N_2O_2$. It is a quinoline derivative having a side chain in the 4th position. It is insoluble in water but soluble in organic solvents. It has an intense bitter taste and kills the microorganisms causing malaria and is used as a medicine against malaria. Quinine is also an antipyretic and lowers the body temperature in high fever.

Quinine

10.4.5 Atropine

Atropine is a tropane alkaloid extracted from the deadly nightshade (*Atropa belladonna*) and other plants of the family *Solanaceae*. It is a secondary metabolite of these plants and serves as a drug with a wide variety of effects. Being potentially deadly, it derives its name from Atropos, one of the three Fates who, according to Greek mythology, chose how a person was to die. Its molecular formula is $C_{17}H_{23}NO_3$ and is made up of tropic acid and tropine. Atropine is a strong poison with bitter taste and causes dilation of the pupils of the eyes.

Atropine

10.4.6 Morphine

Morphine is commonly called opium alkaloid which is extracted from opium poppy, (*Papaver somniferum*). Its molecular formula is $C_{17}H_{19}NO_3$. It has a benzene nucleus, a phenanthrene nucleus and two hydroxyl groups. It is insoluble in water and only slightly soluble in organic solvents. Morphine is used as an analgesic and sedative. It is an extremely powerful opiate analgesic drug and is the principal active agent in opium. It acts directly on the central nervous system to relieve pain. Morphine is highly addictive.

Morphine

10.4.7 Cocaine

Cocaine is a crystalline tropane alkaloid that is obtained from the leaves of the coca plant. It is a stimulant of the central nervous system and an appetite suppressant. Its molecular formula is $C_{17}H_{21}NO_4$. Cocaine in its purest form is a white, pearly product. Cocaine appearing in powder form is a salt, typically cocaine hydrochloride. Cocaine is a potent central nervous system stimulant. Its effects can last from 20 minutes to several hours, depending upon the dosage of cocaine taken, purity, and method of administration. The initial signs of stimulation are hyperactivity, restlessness, increased blood pressure, increased heart rate and euphoria. With excessive dosage the drug can produce hallucinations, tachycardia and itching.

Cocaine

10.4.8 Reserpine

Reserpine is an indole alkaloid antipsychotic and antihypertensive drug. Reserpine was isolated in 1952 from the dried root of *Rauwolfia serpentina* (Indian snakeroot). Reserpine almost irreversibly blocks the uptake (and storage) of noradrenaline and dopamine into synaptic vesicles by inhibiting the Vesicular Monoamine Transporters (VMAT). The molecular formula of reserpine is $C_{33}H_{40}N_2O_9$. Reserpine has a multitude of side-effects, including nausea, vomiting, weight gain, gastric intolerance, gastric ulceration, stomach cramps and diarrhea.

Reserpine

10.4.9 Papaverine

Papaverine is an opium alkaloid used primarily in the treatment of visceral spasm, vasospasm (especially those involving the heart and the brain), and occasionally in the treatment of erectile dysfunction. The molecular formula of papaverine is $C_{20}H_{21}NO_4$. Papaverine is available as a conjugate of hydrochloride, codecarboxylate, adenylate, and teprosylate. The *in vivo* mechanism of action is not entirely clear, but an inhibition of the enzyme phosphodiesterase causing elevation of cyclic AMP levels is significant. It may also alter mitochondrial respiration.

Papaverine

10.5 SUMMARY

- Alkaloids are the basic nitrogenous plant products which have been isolated from roots, leaves, bark and seeds of the plants.
- They are colourless, crytalline and non-volatile compounds which are bitter in taste.
- Alkaloids can be classified into ten different groups on the basis of their chemical structure.
- Alkaloids may be the excretory products of plants and may also have a protective role in plants.
- Alkaloids have various physiological effects in men and other animals.

EXERCISE

1. What are alkaloids? Explain their general properties.
2. Give a brief account of classification of alkaloids.
3. Write short notes on:
    (a) Biological significance of alkaloids
    (b) Coniine
    (c) Nicotine
    (d) Morphine

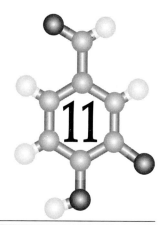

Prostaglandins

Prostaglandins are the chemical compounds which like hormones act as chemical messengers but unlike hormones they act right within the cells where they are synthesized. Prostaglandins were first discovered and isolated from human semen in 1930s by *Ulf Von Euler* of Sweden. He thought that they had come from the prostate gland and hence he named them as Prostaglandins. It has since been determined that they exist and are synthesized virtually in every cell of the body.

Prostaglandins are a class of compounds known as Eicosanoids. Eicosanoids are fatty acid derivatives with a variety of extremely potent hormone like actions on various tissues. They are all derived from the 20 carbon polyunsaturated fatty acid arachidonic acid, from which they take their general name, Greek-'eikosi'-twenty. Eicosanoids are classified into three classes:

(i) Prostaglandins
(ii) Thromboxanes
(iii) Leukotrienes

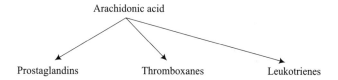

Prostaglandins (PGs) contain a five membered ring of carbon atoms originally part of the chain of arachidonic acid. The basic structural unit is called a prostanoic or prostenoic acid. Each PG differs from the others in the substitution pattern in the cyclopentene ring and the side chains. These differences are responsible for the different biological activities of the members of prostaglandin group.

Based on their cyclopentene/pentene ring substitution pattern, prostaglandins are classified as PGA, PGB, PGC, PGD, PGE, PGF, PGG and PGH. Each general PG class is subclassified on the basis of the degree of unsaturation e.g. PGE is sub classified as PGE_1, PGE_2 and PGE_3. The letters and numbers that follow the initial PG abbreviation indicate the nature of unsaturation and substitution e.g. the subscript 1 in PGE_1 indicates one double bond in the side chains. Similarly in PGE_2 subscript 2 indicates two double bonds in the side chains.

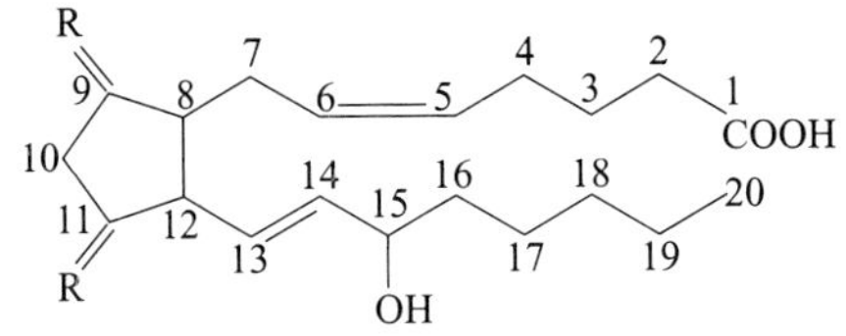

Prostenoic acid

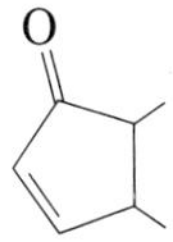

PG A₂   PG B₂   PG C₂

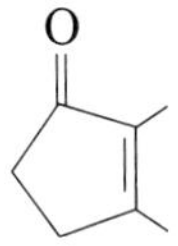

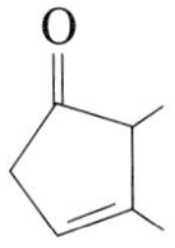

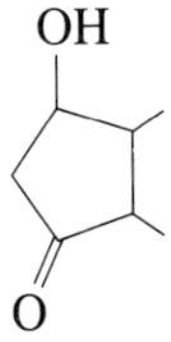

PGD₂   PGE₂   PGF₂

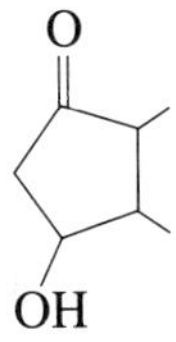

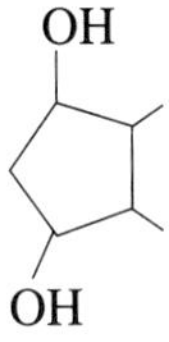

11.1 GENERAL PROPERTIES

Prostaglandins have been detected in almost every mammalian tissue and body fluids. They are produced in minute amounts and their production increases or decreases in response to various stimuli. PGs have broad spectrum and diverse biological effects. They have also been found to modulate cAMP activity in cells either by activating or inhibiting adenyl cyclase activity.

All naturally occurring PGs are 20 C fatty acids containing cyclopentene ring. They have –OH group at 15 position and trans double bond at 13 position.

11.2 OCCURRENCE

PGs were first discovered in seminal plasma and vesicular gland. They have been detected and isolated from pancreas, kidney, brain, thymus, iris, synovial fluid etc. Recently, PGs have been identified in amniotic fluid and umbilical cord vessel.

11.3 BIOSYNTHESIS

The key precursor in biosynthetic pathways of prostaglandins is arachidonic acid. In cells, arachidonic acid is a component of membrane phospholipids such as phosphoinositol. Arachidonic

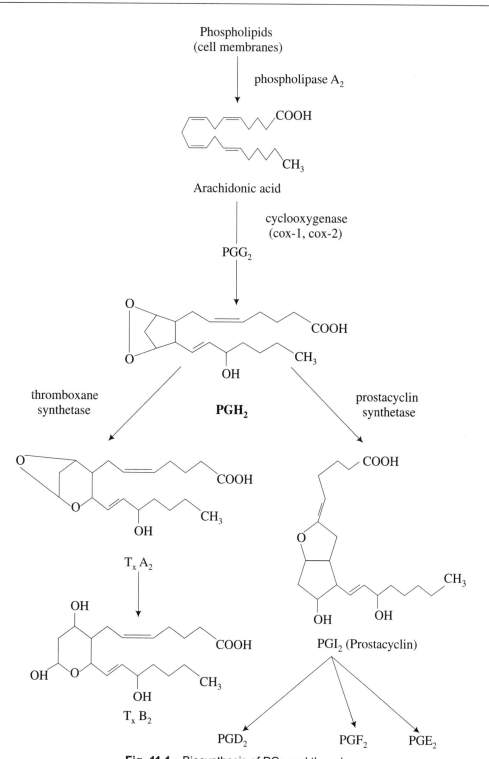

Fig. 11.1 Biosynthesis of PGs and thromboxanes

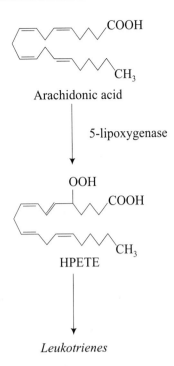

Fig. 11.2 Conversion of arachidonic acid into leukotrienes

acid is liberated from these lipids in response to certain stimuli. The conversion of free arachidonic acid to prostaglandins and other eicosanoids is initiated by oxidative enzymes of cyclooxygenase (PGH synthase) and lipoxygenase families (Fig. 11.1).

First of all cyclooxygenase stereospecifically adds two molecules of oxygen to arachidonic acid to form bicyclic endoperoxide PGG_2. PGG_2 is then reduced by the cyclooxygenase to yield the single 15 (s) – alcohol PGH_2. There exist two different isozymes of cyclooxygenase, a constitutive form (COX – 1) and a highly inducible form (COX – 2). PGH_2 serves as a branch point for the formation of prostacyclin (PGI_2), various prostaglandins as well as thromboxanes. The formation of various derivatives from PGH_2 depends upon the specific tissues, their metabolic capabilities and physiological functions.

Leukotrienes are synthesized from arachidonic acid by the actions of enzymes lipoxygenase which convert them into a variety of acyclic lipid peroxides (hydroperoxyeicosatetraenoic acid HPETEs). HPETEs then produce different types of leukotrienes (Fig. 11.2).

11.4 FUNCTIONS

Prostaglandins play various important roles in the body. They are important mediators of normal physiological events and have been implicated in a variety of pathologies. PGs have been implicated in inflammation, pain, cardio-vascular diseases, cancer, osteoporosis, to name a few.

Depending upon the type of tissue and the nature of the receptors with which PGs interact, the actions of prostaglandins may be stimulatory or inhibitory. Some of the functions of PGs are as follows:

11.4.1 Vasodilators

Prostaglandins are powerful vasodilators. They relax the muscles in the blood vessels and lower the blood pressure. This effect can be local, for example in kidneys wide spread vasodilatation leads to an increase in the flow of blood to the kidney and an increased excretion of salt in the urine.

11.4.2 Effect on Reproduction

Although prostaglandins were first discovered in semen, no specific biological role for them has been defined in male reproductive system. But in females, they play a variety of roles in reproduction. They mediate the control of gonadotrophin releasing hormone over LH secretion, modulate ovulation and stimulate uterine muscle contraction. PGs play important role in inducing labor in pregnant women at term or in inducing therapeutic abortion.

11.4.3 Effect on Immune System

PGEs secreted by macrophages inhibit both T cell and B cell activity. They may also reduce the proliferation of lymphocytes. Prostaglandins, particularly leukotrienes also play important role in inflammatory responses. Leukotrienes enhance inflammation. Many anti-inflammatory drugs act by inhibiting steps in prostaglandin synthetic pathway. Asprin, a well known analegesic, blocks an enzyme cyclooxygenase (COX – 1 and COX –2) which is involved in the conversion of arachidonic acid to prostaglandin, during prostaglandin biosynthetic pathway. By inhibiting or blocking the enzyme cyclooxygenase, the synthesis of prostaglandin is blocked and asprin relieves some of the effects of pain and fever, associated with inflammation.

11.4.4 Effect on Digestive Tract

Prostaglandins affect the functioning of digestive tract. Various PGs may either enhance or inhibit the contraction of smooth muscles of the intestinal wall. They inhibit the secretions of stomach but enhance the bicarbonate and enzyme content of pancreatic juices.

11.4.5 Effect on Blood

PGs also affect haematological responses in the body. PGE diminishes mean corpuscular volume (MCV) and increases erythrocyte deformability. The process of clot formation begins with an aggregation of blood platelets. PGE is a potent inhibitor of human platelet aggregation.

11.5 THROMBOXANES

Thromboxanes are principally synthesized in platelets. Besides platelets, they are also synthesized in neutrophils, lungs, brain, kidney and spleen. They have an oxane ring. They are synthesized from cyclic endoperoxide PGH_2 by the action of enzyme *thromboxane synthetase*. They act in the formation of blood clots and in the reduction of blood flow to the site of a clot.

Table 11.1 Physiological actions of some prostaglandins

S.N.	Prostaglandin	Action
1.	PGD$_2$	– Weak inhibitor of platelet aggregation
2.	PGE$_1$	– Contraction of gastro-intestinal smooth muscle
		– Inhibitor of lipolysis
		– Bronchial vasodilation.
3.	PGE$_2$	– Stimulates uterine smooth muscle relaxation.
		– Protects gastro-intestinal epithelial cells from acid degradation.
		– Promotes inflammation
4.	PGF$_2$	– Leuteolysis.
		– Stimulates uterine smooth muscle contraction.
		– Bronchial constrictor.
5.	PGI$_2$	– Inhibitor of platelet aggregation
		– Uterine relaxant
		– Bronchial dilator.

11.6 LEUKOTRIENES

Leukotrienes is a family of conjugated trienes formed from eicosanoic acids in leukocytes, mast cells and macrophages by lipoxygenase pathway in response to both immunologic and inflammatory stimuli. They are synthesized from arachidonic acid by the addition of hydroxy peroxy groups to arachidonic acid which produces hydroperoxyeicosatetramoates (HPETE). Leukotrienes appear to act as mediators in inflammation and anaphylaxis.

11.7 SUMMARY

• Prostaglandins are the chemical compounds which act as chemical messengers.
• Prostaglandins are the compounds of a class known as eicosanoids.
• They are derived from 20 carbon polyunsaturated fatty acid-arachidonic acid.
• Eicosanoids are classified into three groups-prostaglandins, thromboxanes and leukotrienes.
• Prostaglandins have been detected in most of the mammalian tissues and body fluids. They were first discovered and isolated from human semen by Ulf Von Euler.
• Prostaglandins play wide variety of functions in the body.

EXERCISE

1. What are prostaglandins? Write their general properties.
2. Explain various functions of prostaglandins.
3. Write notes on:
    i. Biosynthesis of prostaglandins
    ii. Thromboxanes

Nucleic Acids

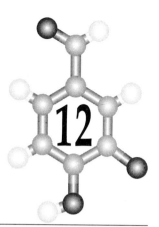

Nucleic acids are the macromolecules present in the cells of all living organisms. They are known as nucleic acids due to their primary occurrence in the nucleus and their acidic nature. P.F. Meischer was the first to discover the nucleic acid associated with the proteins of the cell nucleus of pus cells. In 1932 P.A. Levin has shown that there are two types of nucleic acids – DNA and RNA. DNA (Deoxyribonucleic acid) is a major component of the chromosomes and is the main genetic material in living organisms. Besides nucleus, DNA is also present in chloroplast and mitochondria. RNA (Ribonucleic acid) is present both in the nucleus and in the cytoplasm. In some viruses, RNA is the genetic material.

12.1 STRUCTURE AND COMPOSITION

Both DNA as well as RNA are complex organic molecules. They are the polymers of nucleotides. Nucleotide is the structural unit of the nucleic acids. Nucleotides are themselves complex molecules and have three characteristic components:

  (i) a nitrogenous base,

  (ii) a pentose sugar, and

  (iii) a phosphate.

   Nitrogenous bases are of two types:

  (a) Pyrimidines, and

  (b) Purines.

   Pyrimidines are the single ring nitrogenous bases. They are of three types:

  1. Thymine (T),

  2. Cytosine (C), and

  3. Uracil (U).

   Both thymine (T) and cytosine (C) are present in DNA. In RNA, uracil (U) is present in place of thymine (T) along with cytosine (C).

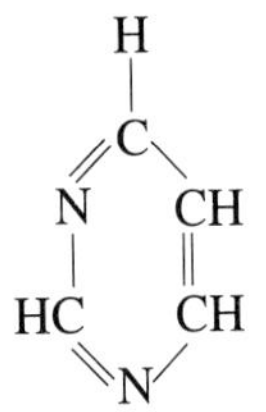

| Pyrimidine | Thymine | Cytosine | Uracil |

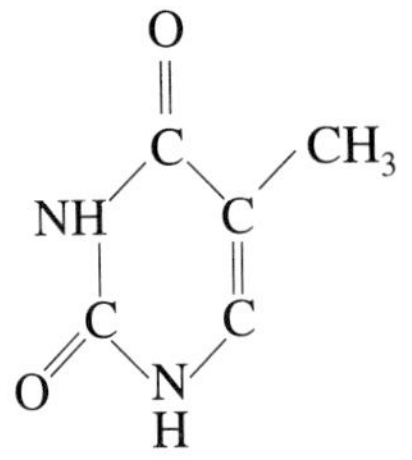

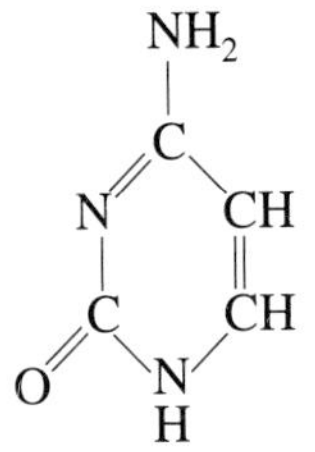

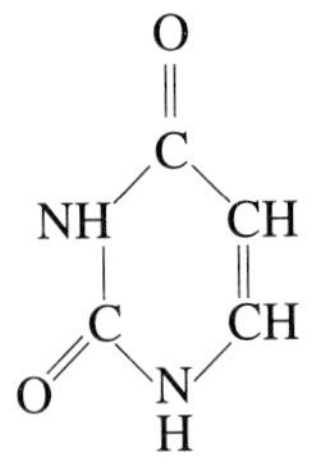

Purines are the double ringed nitrogenous bases. There are two types of purines:

1. Adenine (A), and
2. Guanine (G).

Both of these are present in DNA as well as RNA.

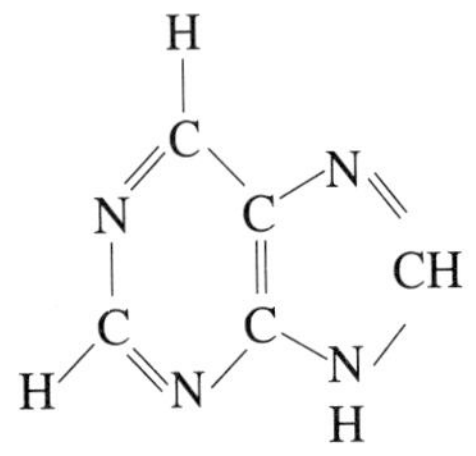

Purine

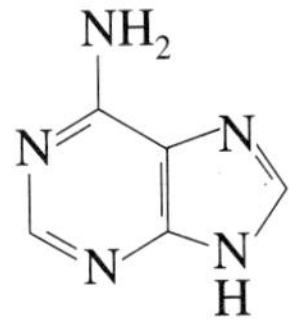

Adenine
(6-amino purine)

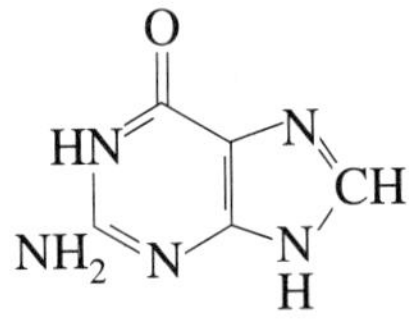

Guanine
(2-amino-6-oxopurine)

Pentose sugars are of two types:

1. Ribose, and
2. Deoxyribose.

Ribose is present in RNA and deoxyribose which lacks oxygen at carbon number two is present in DNA.

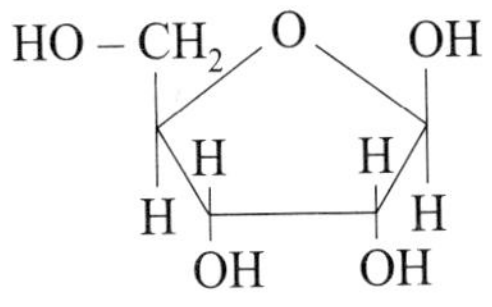

Ribose

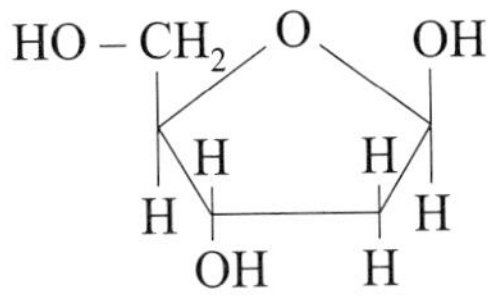

Deoxyribose

12.1.1 Nucleoside and Nucleotides

When one molecule of a nitrogenous base attaches with pentose sugar, it is called *nucleoside*. The pentose sugar may be ribose/deoxyribose.

One molecule of nucleoside combines with one molecule of phosphoric acid to form a *nucleotide*. Nucleotides are the phosphoric esters of nucleosides. Like nucleosides, nucleotides may be of two types on the basis of their pentose sugar:

(i) Ribonucleotides – having ribose sugar and

(ii) Deoxyribonucleotides – having deoxyribose sugar.

Both DNA and RNA are formed by thousands of nucleotides joining together. They are the polymers of nucleotides.

12.2 DNA

DNA, a high molecular weight polymer of nucleotides is present both in prokaryotic and eukaryotic cells. DNA is mainly found in the chromosomes of nucleus. Chromosomes consist of proteins and nucleic acids. Some amount of DNA is also present in some other cell organelles such as mitochondria and chloroplast. DNA is a large molecule with high molecular weight. According to Kuhn (1957), the molecular weight of DNA is 6×10^9 Daltons.

Each DNA molecule is made up of thousands of units of nucleotides. Each nucleotide of DNA is made up of one molecule of nitrogenous base, one molecule of deoxyribose sugar and one molecule of phosphoric acid. In DNA four types of nitrogenous bases are present. They are– adenine, thymine, cytosine and guanine. Out of these adenine and guanine are purine bases i.e. they are double ring compounds, and thymine and cytosine are pyrimidine i.e. single ring compounds. Each nitrogenous base combines with a molecule of deoxyribose sugar to form a nucleoside. Thus there are four types of nucleosides in DNA:

(i) deoxyadenosine – Adenine + deoxyribose sugar

(ii) deoxyguanosine – Guanine + deoxyribose sugar

(iii) deoxycytidine – Cytosine + deoxyribose sugar

(iv) deoxythymidine – Thymine + deoxyribose sugar.

Each nucleoside combines with a phosphoric acid and forms a nucleotide. Thus there are four types of nucleotides present in DNA:

(i) Deoxyadenylic acid

(ii) Deoxyguanylic acid

(iii) Deoxycytidylic acid

(iv) Deoxythymidytic acid

Nucleoside   *Nucleotide*   *Deoxythymidylic acid*

Deoxycytidylic acid

Deoxyadenylic acid

Deoxyguanylic acid

On the basis of their studies Erwin Chargaff et.al. found that the four nucleotides in DNA occur in different ratios in the DNAs of different organisms and the amount of certain bases are closely related. According to Chargaff, "in all DNAs regardless of the species, the number of adenine residues is equal to the number of thymine residues (A = T), and the number of guanine residue is equal to the number of cytosine residues (G = C). Thus, the sum of the purine residues equals the sum of pyrimidine residues". This is sometimes known as "*Chargaff's rule.*"

12.2.1 Molecular Structure of DNA

M.H.F. Wilikins and his associates used X-ray diffraction method to study the structure of DNA. From X-ray diffraction pattern they concluded that the structure of DNA is coiled. In 1953 J.D. Watson and F.H.C. Crick postulated a spiral helix model of DNA. Watson, Crick and Wilikins collectively got Nobel prize in 1962 for this model. According to this model DNA consists of two helical chains coiled around the same axis to form a right handed double helix. The model is known as *double helical model*. In the double helical model the two polypeptide chains run in opposite directions around a common axis. Each chain is formed of many units called nucleotides. Nucleotides are the building blocks of DNA. In nucleotide, the phosphoric acid remains attached with the 5th carbon atom of deoxyribose sugar by ester bond. There is a phosphoester bond between the phosphate molecule of one nucleotide and sugar molecule of other, thus they form phosphate sugar chain. It is called polynucleotide chain. Phosphate molecule is attached with third carbon of deoxyribose sugar. C-3 carbon sugar at one end of the polynucleotide and C-5 carbon at the other end remain free-are not attached with any nucleotide. These ends are known as 3' and 5' end. The two strands of DNA molecule run antiparallel i.e. one strand has phosphodiester bond or linkage in 3' → 5' direction, while in other strand phosphodister linkage is in 5' → 3'.

The nitrogenous bases in DNA lie on the inside of the helix while the sugar-phosphate back bones are on the outside. The adjacent bases are separated by 3.4 A° and the helical structure repeats every 34 A° and thus there are 10 bases (34 A° per repeat/3.4 A° per base) per turn of helix. There is a rotation of 36 degrees per base (360 degree per full turn /10 base per turn). The diameter of the helix is 20 A°. The number of pyrimidine and purine bases is equal in both the polypeptide chains of DNA, that is, the number of A and G is always equal to T and C. The A of one chain can bind only to T of the opposite chain by two hydrogen bonds and C always binds with G with three hydrogen bonds. Thus, both the polypeptide chains are complementary to each other i.e. the sequence of nucleotides of one chain determines the sequence of nucleotides of other chain (Fig. 12.1, 12.2).

The double helical model of Watson and Crick can explain most of the features of DNA.

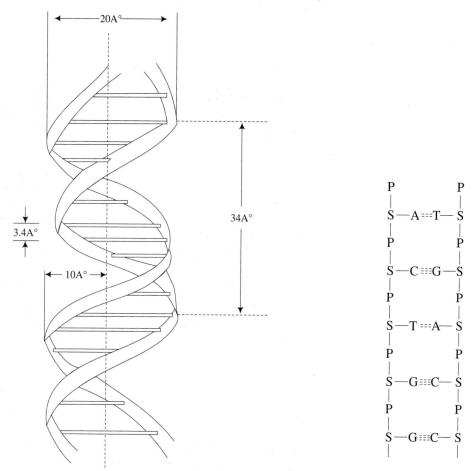

Fig. 12.1 Watson and Crick's model of DNA   **Fig. 12.2** A part of DNA molecule showing base pairing in two polynucleotide chains

12.2.2   Biological Significance

(i) DNA is the genetic material of most of the organisms. The most important function of DNA is to transfer genetic characters from one generation to next generation.

(ii) During protein synthesis, DNA directs the type of protein to be formed. DNA forms mRNA which forms proteins (Fig. 12.3).

(iii) DNA has the power of replication.

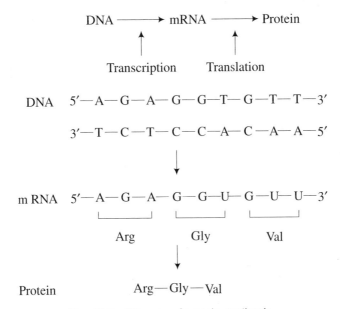

Fig. 12.3   Process of protein synthesis

12.3   STRUCTURAL VARIATIONS IN DNA

DNA is a flexible molecule. It shows some structural variations. On the basis of these structural variations different forms of DNA are present. The biologically most common form of DNA is *B-DNA*. Watson and Crick model of DNA is known as B-DNA. Other major structural variants are *A-DNA* and *Z-DNA*.

12.3.1   A-DNA

A-DNA is present under the conditions of dehydration. Under dehydrating conditions B-DNA undergoes a reversible conformatinal change to form A-DNA. Like B-DNA, A-DNA is also a right handed double helix made up of antiparallel strands but A-helix is wider and shorter than the B-helix. The number of base pairs per helical turn is 11 with a pitch of 28 A°. This gives it an axial hole. The plane of the base pairs in A-DNA is tilted about 20° with respect to the helix axis. Due to these structural changes A-DNA has a deep major groove and a very shallow minor groove.

The A-helix is not confined to dehydrated DNA. Double stranded regions of RNA and some RNA-DNA hybrids also adopt a form very similar to that of A-DNA.

12.3.2 Z-DNA

Z-DNA, a third type of DNA helix was discovered by Alexander Rich and his associates. Z-DNA is a left handed double helix, in contrast to the right handed A and B-helices. The number of base pairs per turn is 12 with a pitch of 45 A°. The structure appears more slender and elongated. In Z-DNA the major groove is not apparent but the minor groove is narrow and deep. The DNA back bone takes on a zig-zag appearance and hence this is called Z-DNA. The Z-form is adopted by short oligonucleotides with alternating sequences of purines and pyrimidines at high salt concentrations. High salt concentrations are required to minimize electrostatic repulsion between closest approaching phosphate groups on opposite strands. The biological role of Z-DNA is not yet defined but it may play a role in regulating the expression of certain genes or in genetic recombinations.

Comparison of A-, B- and Z-DNA.

	A-DNA	B-DNA	Z-DNA
1. Helical sense	Right handed	Right Handed	Left handed
2. Helix diameter	26 A°	20 A°	18 A°
3. Base pairs per turn of helix	11	10	12
4. Helix pitch	28 A°	34 A°	45 A°
5. Rise per base pair	2.6 A°	3.4 A°	3.7 A°
6. Tilt of base pairs from normal to the helix axis	20°	6°	7°
7. Major groove	Narrow and deep	Wide and quite deep	Flat
8. Minor groove	Broad and shallow	Narrow and quite deep	Narrow and deep.
9. Glycosidic bond	anti	anti	anti for pyrimidine and syn for purines.

12.4 RNA

RNA is generally found in the cytoplasm and the nucleolus. Besides this, it is also found in mitochondria and chloroplast. RNA is made up of long unbranched polynucleotide chain of single strand. The single strand coils itself to form a helix. Like DNA, the polynucleotide chain of RNA is made up of thousands of nucleotides linked together by $3' - 5'$ phosphodiester bonds forming linear sequence. In RNA the sugar is ribose sugar. Ribose sugar attaches with nitrogenous base and forms a nucleoside. The nucleosides are called ribonucleosides. One molecule of nucleoside combines with one molecule of phosphoric acid to form a nucleotide. Nucleotides of RNA are called ribonucleotides.

In RNA both types of nitrogenous bases–pyrimidine and purines are present. The purines are adenine and guanine, as in DNA. The pyrimidine bases in RNA are cytosine and uracil. In RNA unlike DNA, thymine is replaced by uracil (U).

Thus, four types of nucleotides present in RNA are:
 (i) Adenylic acid = Adenosine + phosphoric acid
 (ii) Guanylic acid = Guanosine + phosphoric acid
(iii) Cytidylic acid = Cytosine + phosphoric acid
(iv) Uridylic acid = Uracil + phosphoric acid

12.5 TYPES OF RNA

Three types of RNA are present in the cells. They are:
  (i) Messenger RNA (mRNA),
  (ii) Ribosomal RNA (rRNA), and
  (iii) Transfer RNA (tRNA).

12.5.1 Messenger RNA (mRNA)

Jacob and Monod proposed the term mRNA for the portion of cellular RNA carrying information for protein synthesis from DNA to the sites of protein synthesis. The process of protein synthesis can be divided into two steps:
  (i) the formation of mRNA from DNA (transcription) and
  (ii) the formation of proteins from mRNA (translation).

Since proteins occur in different sizes, mRNA varies greatly in length and molecular weight. In prokaryote a single mRNA may code one or more polypeptide chains. If mRNA codes for a single polypeptide chain, it is called *monocistronic* and if it codes for two or more different polypeptide chains, it is called *polycistronic*. The molecular weight of an average sized mRNA is about 500,000 D and sedimentation coefficient is 85. It forms 5% to 10% of the total RNA present in the cell.

mRNA is formed as a complementary strand from one of the strands of DNA, the base sequence of each mRNA is complementary to that of DNA.

In eukaryotic the mRNA has the following structural features:

1. *Cap*: A cap is found at the 5′ end of mRNA in which methylated guinine is present. The rate of protein synthesis depends upon the presence of cap, without the cap the mRNA molecule binds very poorly to the ribosomes.

2. *Non-coding region-1 (NC-1)*: The cap is followed by a region of 10 – 100 nucleotides which is rich in A and U residues. This portion does not translate protein.

3. *Initiation Codon*: In both prokaryotes and eukaryotes the initiation codon is AUG.

4. *Coding region*: This region translates protein.

5. *Termination*: Termination of translation on mRNA is carried by termination codon. Termination codons are UAA, UAG and UGA.

6. *Non-coding region-2 (NC-2)*: This region consists of 50–150 nucleotides and does not translate protein.

7. *Poly [A] Sequence*: At 3′ end poly A sequence is found. It initially consists of 200–250 nucleotides but becomes shorter with age (Fig. 12.4).

12.5.2 Ribosomal RNA (rRNA)

rRNA is found in ribosomes. It forms 80% of the total RNA of the cell. rRNA molecules may be short compact rod, a compact coil or extended strand depending upon ionic strength, temperature and pH. rRNA consists of a single strand and the same strand twisted upon itself in some regions and forms helix. The helical regions are connected by interlinked single strand region. In the helix

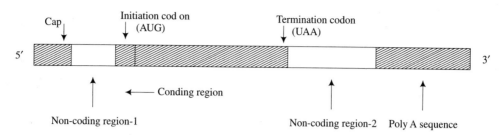

Fig. 12.4 General features of eukaryotic mRNA

region most of the base pairs are complementary and are joined with hydrogen bonds. In the unfolded single strand region the base pairs are not complementary. Hence rRNA does not show purines and pyrimidines equality. rRNA differs in base contents from tRNA and mRNA. The quantity of guanine and cytosine is relatively much more (Fig. 12.5).

Types of rRNA

On the basis of sedimentation and molecular weight, rRNA are of the following three types:

   (i) High molecular weight rRNA with molecular weight more than a million (21–29S rRNA).

  (ii) High molecular weight rRNA with molecular weight less than a million (12–18S rRNA).

 (iii) Low molecular weight rRNA with molecular weight approximately 40,000 (5S RNA).

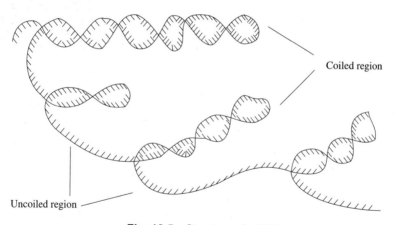

Fig. 12.5 Structure of r RNA

12.5.3   Transfer RNA (tRNA)

tRNA molecules are found in the cytoplasm of the cell. tRNA comprises 10–15% of the total weight of RNA present in the cell. Its molecular weight is 25,000–30,000 D. They bind to 20 different amino acids and take them to the sites of protein synthesis. In the cytoplasm of the cell about 20 different types of tRNA molecules are present which are quite similar to each other.

A tRNA molecule is made up of 75 – 80 nucleotides and is the smallest of all the nucleic acids. Each molecule of tRNA coils itself in such a manner that its both arms are coiled around each other. Both ends of its polyncleotide chains show identical structure. At 3′end of chain CCA (cytidylic acid, cytidylic acid and adenylic acid) are present and 5′ end terminates in G (guanylic acid). At the turn of each tRNA molecule chain there is a definite sequence of three nitrogenous bases which forms *anticodon*.

On the basis of nucleotide-analysis of the RNA of yeast, Robert Holey proposed a *Clover-Leaf model* of tRNA. According to this model tRNA has four arms out of which one arm has anticodon (Fig. 12.6). On each tRNA molecule following four special sites are present:

(i) *Amino acid attachment site*: It is situated at 3′ end of tRNA chain. The – OH group joins with any specific amino acid to form aminoacyl-tRNA in the presence of ATP.

(ii) *Recognition site*: tRNA attaches with specific amino acid at this site.

(iii) *Anticodon (codon recognition site)*: The base sequence of codon recognition site is complementary of the triplet situated on mRNA. This site is very important for tRNA sequence because besides deciding specific codon, it also determines the pairing of tRNA.

(iv) *Ribosome recognition site*: During protein synthesis, tRNA attaches with ribosome at this site.

In tRNA, besides normal nitrogenous bases – C, G, A and uracil, a few abnormal bases such as pseudouridine, isosinic acid, methyl guanine etc. are also present. The presence of these bases does not create any hindrance of any type of pairing of tRNA with mRNA. The presence of these abnormal bases may possibly stop intramolecular base pairing of tRNA or they help in *amino-acyl tRNA synthatase* enzyme recognition.

12.6 COMPARISON BETWEEN DNA AND RNA

	DNA	RNA
1.	It is known as deoxyribonucleic acid.	It is known as ribonucleic acid.
2.	It contains deoxyribose sugar.	It contains ribose sugar.
3.	Its molecule is made up of double stranded helix.	Its molecule is made up of single strand.
4.	The basic unit of DNA is a nucleotide. A nucleotide consists of nucleoside and one molecule of phosphoric acid. A nucleoside consists of a nitrogenous base and a sugar.	In RNA also the basic unit is nucleotide which consists of nucleoside and a molecule of phosphoric acid.
5.	There are four types of nitrogenous bases present in DNA: Purines-adenine and guanine Pyrimidines-*thymine* and cytosine.	Four types of nitrogenous bases are also present in RNA: Purines-adenine and guanine Pyrimidines-*uracil* and cytosine.
6.	The base composition is A/T = G/C = 1.	It is not equal in RNA.
7.	DNA is the genetic material and controls all the activities of the cell.	It plays important role in protein synthesis.
8.	DNA consists of a large number of nucleotides and its molecular weight is more.	The molecular weight of RNA is comparatively less.
9.	DNA is mainly present in nucleus. Besides nucleus, it is also present in mitochondria and chloroplast.	RNA is present in nucleus-nucleoplasm, nucleolus and cytoplasm.

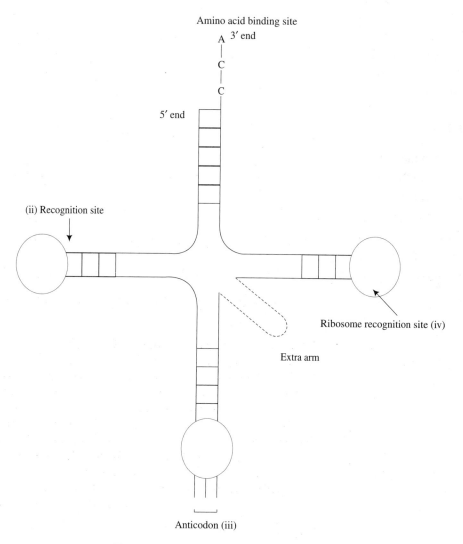

Fig. 12.6 Basic plan of the structure of tRNA

12.7 SUMMARY

- DNA and RNA are the nucleic acids present in the nucleus of the cells.
- They are the polymers of nucleotides.
- A nucleotide is composed of a nitrogenous base, a pentose sugar and a phosphate molecule.
- The pentose sugar present in DNA is deoxyribose.
- Nitrogenous bases present in DNA are—adenine, thymine, cytosine and guanine.
- To explain the structure of DNA, Watson and Crick proposed a model in 1953, known as double helical model.

- Three structural variants are present in DNA–B-DNA, Z-DNA and A-DNA.
- Pentose sugar present in RNA is ribose.
- There are three types of RNA–mRNA, rRNA and tRNA.
- DNA is the hereditary material of the organisms. RNA mainly participates in the protein synthesis.

EXERCISE

1. Explain the Watson-Crick model of DNA.
2. Compare RNA with DNA.
3. Differentiate between:
   (i)   Nucleotide and Nucleoside
   (ii) Pyrimidine and Purine
   (iii) B-DNA and Z-DNA
4. Explain the structure and functions of tRNA.

Part II
Bioenergetics and Metabolism

Metabolism of Carbohydrates

Carbohydrates are the major source of energy in the organisms. They are the main components of the food. Dietary carbohydrates mainly consist of polysaccharides (starch) and disaccharides (sucrose, lactose and maltose). These are hydrolyzed into monosaccharides and are absorbed in the blood. Glucose is an important monosaccharide which serves as the major metabolic fuel of the cells and tissues. Non-glucose monosaccharides are also converted into glucose. Glucose is metabolised by different ways to produce energy in the body. The main pathways of carbohydrate metabolism are:

  (i) Glycolysis,
 (ii) Citric acid cycle,
(iii) Glycogenesis,
(iv) Glycogenolysis,
 (v) Gluconeogenesis, and
(vi) Phosphate pentose pathway.

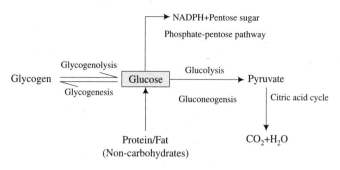

Fig. 13.1 Metabolism of carbohydrates

13.1 GLYCOLYSIS

Glycolysis is a process in which one molecule of glucose is converted into two molecules of three carbon compound pyruvate. The term glycolysis is derived from two Greek words-*glykys*, meaning sweet (sugar) and *lysis*, meaning splitting. The pathway is also known as *Embden-Meyerhof-Parnas pathway*. Glycolysis occurs virtually in all tissues. Enzymes involved in glycolysis are extra mitochondrial and hence glycolysis occurs in the cytoplasm of the cells. Glycolysis can be divided into two stages. In the first stage, one molecule of glucose is converted into two molecules of glyceraldehyde-3-phosphate. The process requires two molecules of ATP.

In the second stage the molecules of glyceraldehyde-3-phosphate are converted into pyruvate with the release of 4ATP and 2 NADH molecules (Fig. 13.2).

$$\boxed{\text{(1)}\ \text{Glucose}}$$
$$\downarrow\ 2\text{ATP}$$
$$\text{(2) Glyceraldehyde-3-phosphate}$$
$$\downarrow\ 4\text{ATP}+2\text{NADH}$$
$$\text{(2)}\ \boxed{\text{Pyruvate}}$$

Fig. 13.2 Pathway of glycolysis

The whole process occurs in ten reactions (Fig. 13.3).

The reactions of glycolysis are as follows:

13.1.1 Reactions of Glycolysis

(i) *Phosphorylation of glucose*: In the first step of glycolysis glucose is phosphorylated at C-6 to yield *glucose-6-phosphate* with ATP as the phosphoryl donor. The reaction is catalyzed by the enzyme *hexokinase*. A kinase is an enzyme that transfers phosphoryl groups between ATP and a metabolite. Hexokinase is a relatively non-specific enzyme that catalyzes the phosphorylation of hexoses such as D-glucose, D-fructose and D-mannose. Hexokinase requires Mg^{2+} for its activity as the true substrate for the enzyme is not ATP^{4-} but $Mg\ ATP^{2-}$ complex. This enzyme is present in all cells of all organisms. Liver cells also contain *glucokinase* which is a specific enzyme.

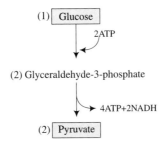

Glucose   Glucose-6-phosphate

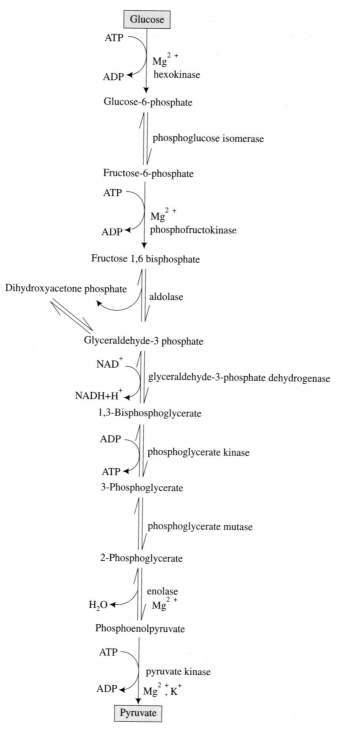

Fig. 13.3   Embden-Meyerhof pathway of glycolysis

(ii) *Formation of fructose-6-phosphate*: Glucose-6-phosphate is converted to *fructose-6-phosphate* by the enzyme *phosphohexose isomerase* (phosphoglucose isomerase). In this reaction there is isomerization of an aldose to a ketose.

Glucose-6-phosphate   Fructose-6-phosphate

(iii) *Phosphorylation of fructose-6-phosphate to fructose 1,6 bisphosphate*: Fructose-6-phosphate is phosphorylated by the enzyme *phosphofructokinase* to yield *fructose 1, 6 bisphosphate*.

Fructose-6-phosphate   Fructose 1, 6 bisphosphate

The product is fructose 1, 6 bisphosphate rather than a diphosphate since two phosphate groups are not attached directly to each other.

(iv) *Cleavage of fructose 1, 6 bisphosphate*: Fructose 1, 6 bisphosphate is cleaved into two trioses *glyceraldehyde 3-phosphate* and *dihydroxyacetone phosphate*. The reaction is catalyzed by the enzyme *aldolase*.

Fructose 1,6 bisphoshate   Dihydroxyacetone phosphate   Glyceraldehyde-3-phoshate

(v) *Isomerization*: Only glyceraldehyde-3-phosphate continues along the glycolytic pathway. Dihydroxyacetone phosphate can be readily converted into glyceraldehyde -3-phosphate as both are isomers. This isomerization is catalyzed by the enzyme *triose phosphate isomerase*. As a result of this isomerization two molecules of glyceraldehyde-3-phosphate are formed from one molecule of fructose 1, 6 bisphosphate.

Dihydroxyacetone phosphate   Glyceraldehyde-3-phosphate

(vi) *Oxidation and phosphorylation of glyceraldehyde-3-phosphate*: Glyceraldehyde-3-phosphate is oxidized and phosphorylated simultaneously in the presence of NAD^+ and Pi to form *1, 3 bisphosphoglycerate*. The reaction is catalyzed by the enzyme *glyceraldehyde 3-phosphate dehydrogenase*. In this reaction NAD^+ is reduced to form NADH.

Glyceraldehyde-3-phosphate Inorganic phosphate   1,3 Bisphosphoglycerate

(vii) *Formation of 3-phosphoglycerate from 1, 3 bisphosphoglycerate*: 1, 3 Bisphosphoglycerate is converted into *3-phosphoglycerate* with the formation of ATP. The reaction is catalyzed by the enzyme *phosphoglycerate kinase* which transfers the high energy phosphoryl group from the carboxyl group of 1, 3 bisphosphateglycerate to ADP forming ATP and 3-phosphoglycerate.

1,3 Bisphosphoglycerate   3-Phosphoglycerate

(viii) *Conversion of 3-phosphoglycerate to 2-phosphoglycerate*: 3-Phosphoglycerate is converted to *2-phosphoglycerate* by the enzyme *phosphoglycerate mutase*. A mutase is an enzyme which catalyzes the transfer of a functional group from one position to another on a molecule.

3-Phosphoglycerate → (phosphoglycerate mutase) → 2-Phosphoglycerate

(ix) *Dehydration of 2-phosphoglycerate to phosphoenolpyruvate*: 2-Phosphoglycerate is dehydrated to *phosphoenolpyruvate* by the enzyme *enolase*.

2-Phosphoglycerate → (enolase) → Phosphoenolpyruvate + H_2O

(x) *Formation of pyruvate*: In the final reaction of glycolysis, the phosphoryl group from phosphoenolpyruvate is transferred to ADP with the formation of ATP and *pyruvate*. The reaction is catalyzed by *pyruvate kinase* which requires both K^+ and either Mg^{2+} or Mn^{2+}.

Phosphoenolpyruvate + ADP → (pyruvate kinase, K^+, Mg^{2+}) → Pyruvate + ATP

13.1.2 Energy Yield in Glycolysis

Glycolysis results in the formation of two molecules of pyruvate from one molecule of glucose. One molecule of glucose is first converted into two molecules of glyceraldehyde-3-phosphate. This process requires 2 ATP molecules. Oxidation of each molecule of glyceraldehyde 3-phosphate to a pryuvate yields 2 ATP and 1 NADH. Oxidation of two molecules of glyceraldehyde -3-phosphate yields 4 ATP and 2 NADH. Each molecule of NADH is reoxidized to NAD^+ with the formation of 3 molecules of ATP. Thus, as a result of glycolysis 10 ATP are formed. 2 ATP are consumed during the conversion of glucose into glyceraldehyde-3-phosphate. Hence the net gain is 8 ATP. Energy yield during glycolysis can be summerized as:

Reaction	*ATP used/produced*
* Glucose to Glucose 6-phosphate	– 1
* Fructose-6-phosphate to fructose-1, 6-bisphosphate	– 1
* 1,3 Bisphosphoglycerate to 3-phosphoglycerate	+ 2
* Phosphoenolpyruvate to pyruvate	+ 2
* Glyceraldehyde-3-phosphate to 1, 3 bisphosphoglycerate (2NADH = 6 ATP)	+ 6
NET GAIN	8

13.2 FATE OF PYRUVATE

The pyruvate formed, as a result of glycolysis, can be further metabolised via any of the three catabolic pathways (Fig. 13.4):

(i) Under aerobic conditions, pyruvate is oxidized completely to yield CO_2 and H_2O by Citric Acid cycle.

(ii) Under anaerobic conditions, in some microorganisms and under hypoxia in skeletal muscles, pyruvate is reduced to lactate via lactic acid fermentation.

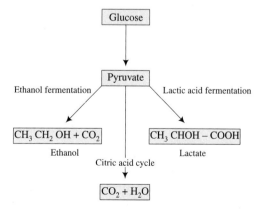

Fig. 13.4   Fate of pyruvate

(iii) In some plant tissues, invertebrates and microorganisms, under hypoxic or anaerobic conditions, pyruvate is converted into ethanol and CO_2 via ethanol (alcohol) fermentation.

Ethanol fermentation: In yeast and other microorganisms, pyruvate is fermented to ethanol and CO_2. The reaction occurs in two steps: First, pyruvate is decarboxylated to form acetaldehyde and CO_2. The reaction is catalyzed by *pyruvate decarboxylase*. In the second step, acetaldehyde is reduced to ethanol by NADH. The reaction is catalyzed by *alcohol dehydrogenase*.

Lactic acid fermentation: When oxygen supply is not sufficient e.g. during vigorous exercise, pyruvate is reduced to lactate to regenerate NAD^+ from NADH. The reaction is catalyzed by *lactate dehydrogenase*.

Formation of Acetyl-CoA: Under aerobic conditions, pyruvate undergoes oxidative decarboxylation to form acetyl-CoA. The reaction is catalyzed by *pyruvate dehydrogenase complex*-a multi enzyme complex.The enzyme pyruvate dehydrogenase complex contains multiple copies of three enzymes-*pyruvate dehydrogenase, dihydrolipoyl transacetylase* and *dihydrolipoyl dehydrogenase* and requires five coenzymes for its activity – thiamin pyrophosphate (TPP), lipoic acid, Co-ASH, FAD and NAD^+.

13.3 CITRIC ACID CYCLE

Under aerobic conditions pyruvate is completely oxidized to yield CO_2 and H_2O. Pyruvate is first converted to acetyl-CoA which is then oxidized to produce CO_2 and H_2O (Fig. 13.5). The whole

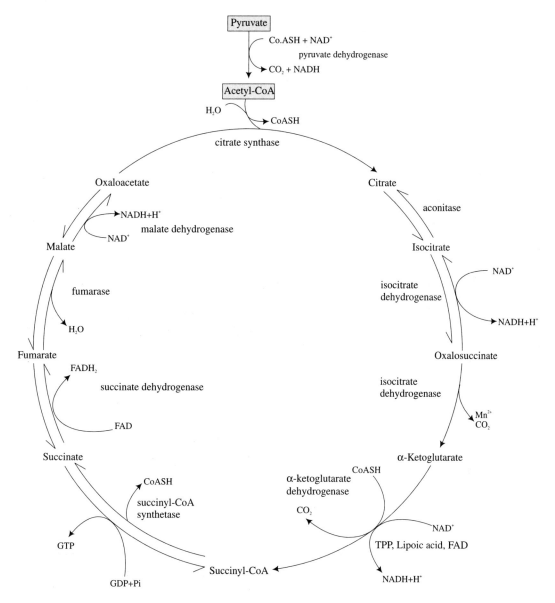

Fig. 13.5 Citric acid cycle

process occurs in a cyclic manner and is called *Citric Acid Cycle*. The cycle is also known as *TCA Cycle* (Tricarboxylic Acid Cycle) or *Krebs Cycle*, after its discoverer Hans Krebs. Acetyl-CoA is derived not only from the oxidation of carbohydrates but also from β-oxidation of fatty acids and partly from certain amino acids. Thus, TCA cycle is the central pathway for recovering energy from several metabolic fuels. The enzymes for TCA cycle are located in mitochondrial matrix.

13.3.1 Reactions of TCA Cycle

The whole TCA cycle involves eight reactions:

(i) *Formation of citrate*: The first reaction of TCA cycle is the condensation of oxaloacetate and acetyl-CoA to yield *citrate* and CoA. The reaction is catalyzed by *citrate synthase*. A H_2O molecule is required to hydrolyze the high energy bond linkage between the acetyl group and CoA. Co A-SH released is reutilized for oxidative decarboxylation of pyruvate.

Acetyl-Co A.   Oxaloacetate   Citrate

(ii) *Formation of isocitrate*: Citrate is reversibly isomerized to *isocitrate* by the enzyme *aconitase* with *cis-aconitate* as the intermediate. The isomerization of citrate is accomplished by dehydration step followed by a hydration step.

Citrate   Cis-aconitate   Isocitrate

(iii) *Formation of α-ketoglutarate*: Oxidative decarboxylation of isocitrate results in the formation of *α-ketoglutarate*. The reaction is catalyzed by the enzyme *isocitrate dehydrogenase*. Oxalosuccinate is formed as an intermediate.

Isocitrate   Oxalosuccinate   α-Ketoglutarate

(iv) *Oxidative decarboxylation of α-ketoglutarate*: α-ketoglutarate undergoes oxidative decarboxylation of form *succinyl-CoA*. *α-Ketoglutarate dehydrogenase complex* catalyzes the reaction. NAD^+ serves as the electron acceptor. *α-Ketoglutarate dehydrognase complex* closely resembles the *pyruvate dehydrogenase complex* both in structure and function and requires identical coenzymes and cofactors.

(v) *Formation of succinate*: Succinyl-CoA is converted to *succinate* by *succinate thiokinase* (also known as *succinyl-CoA synthetase*). The reaction is reversible. GTP is formed in this reaction which may be converted to ATP in the presence of enzyme *nucleoside diphosphate kinase*.

(vi) *Oxidation of succinate to fumarate*: Succinate is oxidized to *fumarate* by the enzyme *succinate dehydrogenase*. FAD is the electron acceptor in this reaction. The enzyme *succinate dehydrogenase* is inhibited by malonate which acts as a *competitive inhibitor*.

(vii) *Formation of malate*: *Fumarase* catalyzes the hydration of fumarate to form malate. This is a reversible reaction.

Fumarate   Malate

(viii) *Formation of oxaloacetate*: This is the last reaction of citric acid cycle in which malate is oxidized by the enzyme *malate dehydrogenase* to form *oxaloacetate*. NAD^+ acts as the H acceptor.

Malate   Oxaloacetate

Oxaloacetate combines with a fresh molecule of acetyl-CoA and the whole process is repeated.

13.3.2 Energetics of TCA Cycle

During TCA cycle each molecule of acetyl-CoA undergoes oxidation to yield 3 molecules of NADH (one each, at conversion of isocitrate to α-ketoglutarate; α-ketoglutarate to succinyl – CoA; malate to oxaloacetate), one molecule of $FADH_2$ (succinate to fumarate), one GTP (succinyl – CoA to succinate). Each molecule of NADH undergoes oxidation by electron transport chain to yield 3 molecules of ATP. Similarly, $FADH_2$ undergoes oxidation to yield 2 ATP. Thus, during oxidation of one molecule of acetyl – CoA, 12 ATP are formed (3 NADH = 9 ATP; 1 $FADH_2$ = 2 ATP And 1 GTP = 1 ATP).

We have seen during glycolysis that each molecule of glucose yields two molecules of pyruvate and 8 ATP are gained. Hence, when these two molecules of pyruvate undergo complete oxidation by TCA cycle 24 ATP are formed. In addition, 2 NADH are formed during the formation of acetyl-CoA by pyruvate through dehydrogenase complex which yield 6 ATP.

Thus, complete oxidation of one molecule of glucose will yield 38 ATP molecules (glycolysis 8, pyruvate dehydrogenase complex 6, TCA cycle 24).

13.4 GLYCOGENESIS

Glycogenesis is the process of formation of glycogen from glucose. The major sites of glycogenesis are liver and muscles but it can occur in every tissue in the body to some extent. Glycogenesis is a very important process as excess of glucose is converted to glycogen and is stored in this form for utilization at the time of requirement (Fig. 13.6).

Glucose

↓

Glucose-6-phosphate

↓

Glucose-1-phosphate

⌐UTP

→Pi

UDP-Glucose

glycogen synthase

→ UDP

Branching enzyme

Glycogen

Fig. 13.6   Pathway of glycogenesis

Glycogenesis begins with the phosphorylation of glucose to *glucose-6-phosphate* by the enzyme *glucokinase (hexokinase)* in the presence of ATP and Mg^{2+}.

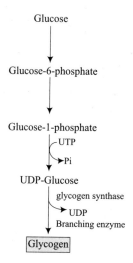

Glucose-6-phosphate is converted to *glucose-l-phosphate* by the enzyme *phosphoglucomutase.*

Glucose-l-phosphate reacts with UTP to form *UDP-G (uridine diphosphoglucose).* The reaction is catalyzed by *UDP-G pyrophosphorylase* with the elimination of a molecule of pyrophosphate.

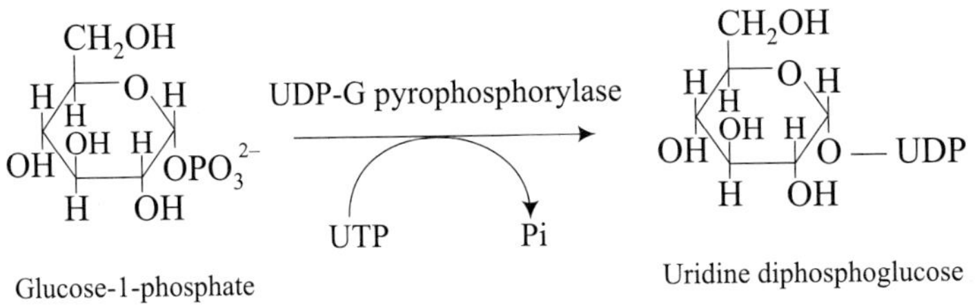

Glucose-1-phosphate — UDP-G pyrophosphorylase → Uridine diphosphoglucose

For synthesis of glycogen, a pre-existing glycogen chain (glycogen primer) is required. UDP-G transfers the glucose molecule to a pre-existing glycogen primer. C-1 of the glucose of UDP-G forms a glycosidic bond with the C-4 of a terminal residue of glycogen primer in the presence of enzyme *glycogen synthase*. UDP is liberated in the process.

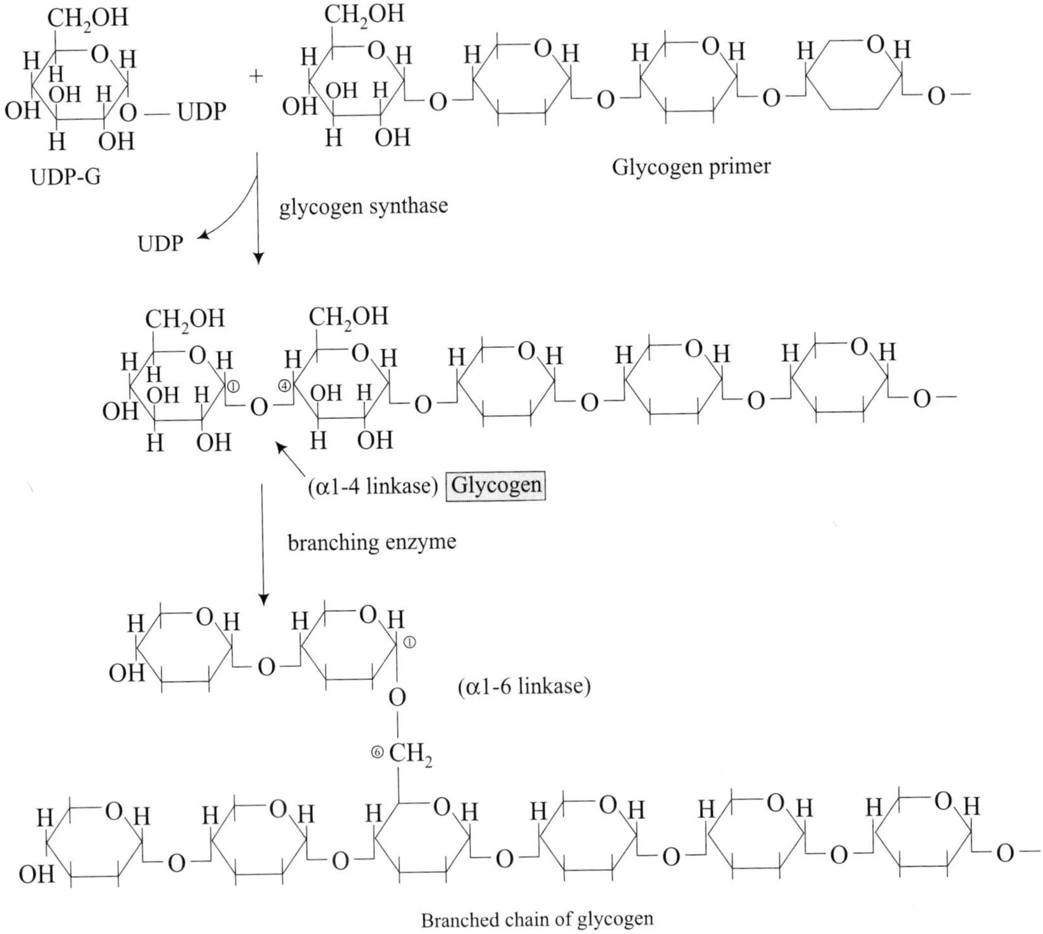

Fig. 13.7 Formation of glycogen

In this way an existing glycogen chain is repeatedly extended by one glucose unit at a time by the successive α1-4 linkage. Glycogen synthase is the principal enzyme which regulates glycogen formation. Glycogen synthase can add glycosyl residue only if polysaccharide chain already contains more than four residues. When the chain has become minimum of 11 glucose residues, another enzyme *branching enzyme* transfers a part of α1-4 chain (at least 6 glucose residues) to a neighbouring chain to form α1-6 linkage, thus forming a branching point in the molecule. The branches now grow further by further addition of α1-4 glycosyl units and further branching (Fig. 13.7).

The glycogen formed is stored in liver and muscles.

13.5 GLYCOGENOLYSIS

The break down of glycogen to glucose is called glycogenolysis. When the blood glucose level falls, glycogen is broken down to glucose to maintain the normal level of blood glucose.

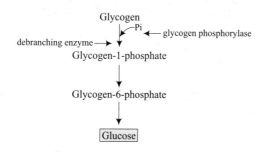

Fig. 13.8 Pathway of glycogenolysis

The first step in the breakdown of glycogen is catalyzed by the enzyme *glycogen phosphorylase* which catalyzes the cleavage of a terminal α1-4 linkage of glycogen to remove one glucose residue as *glucose-1-phosphate*. The removal of glucose residues continues until about four to five glucose residues remain on either side of the α1-6 branch. The glycogen branch with four glycogen residues is called a 'limit branch.'

Enzyme phosphorylase exists in two forms – 'active' and 'inactive' form in both liver and muscle. These two forms are interconvertible. Muscle phosphorylase is immunologically distinct from liver phosphorylase and is not affected by glucagon.

Now, another enzyme *glycosyl transferase* transfers a block of three glycosyl residues from a limit branch of glycogen to non-reducing end of another branch. This exposes a single glucose residue joined by an α1-6 glycosidic linkage. Another enzyme *α–1→6 glucosidase* hydrolyzes the α–1→6 glycosidic bond, resulting in the release of a free glucose molecule. *Glycogen de-branching enzyme* acts both as a transferase and a glucosidase enzyme (Fig. 13.9).

Glucose-1-phosphate formed on the phosphorylatic cleavage of glycogen is converted into *glucose-6-phosphate* by *phosphoglucomutase*.

In liver, a hydrolytic enzyme *glucose-6-phosphatase* cleaves the phosphoryl group from glucose-6-phosphate to form free glucose and orthophosphate. In muscles, enzyme glucose-6-phosphatase is absent and hence glucose-6-phosphate enters into glycolytic cycle.

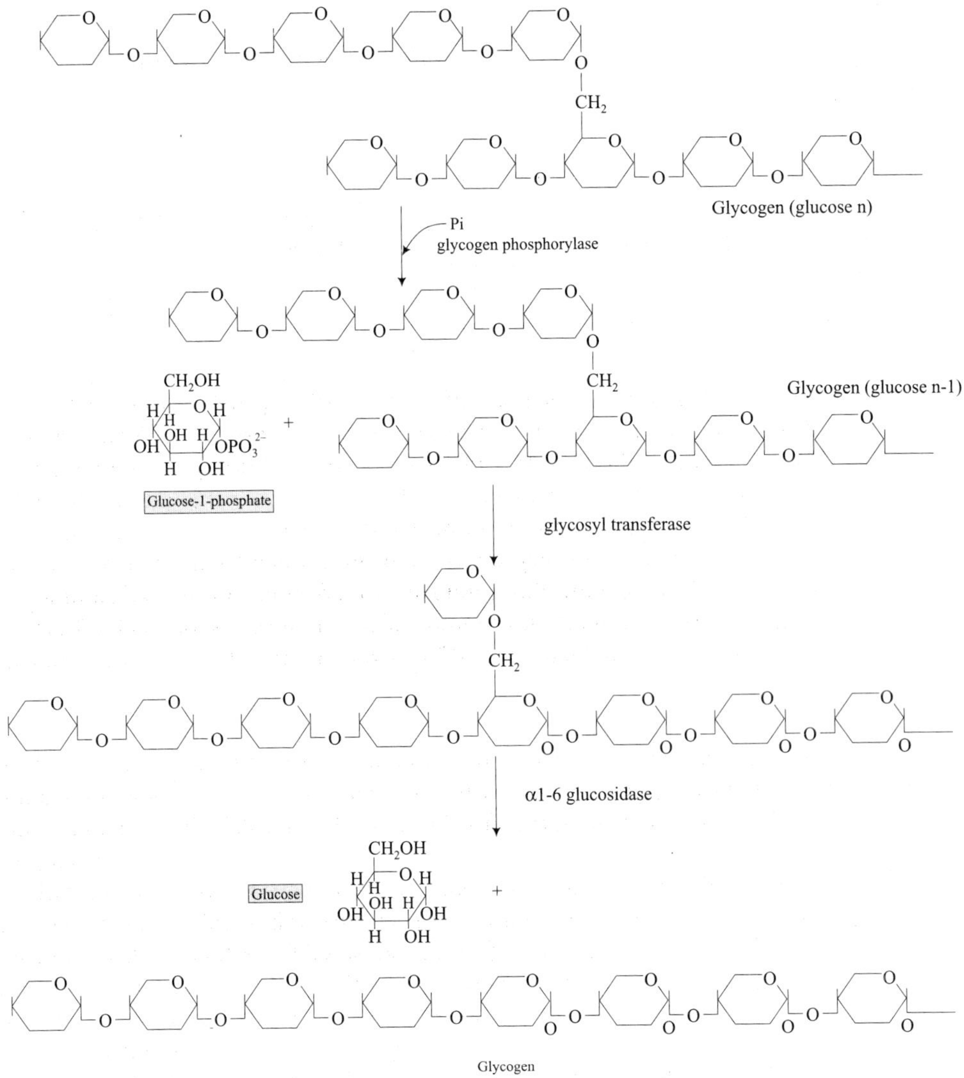

Fig. 13.9 Glycogenolysis

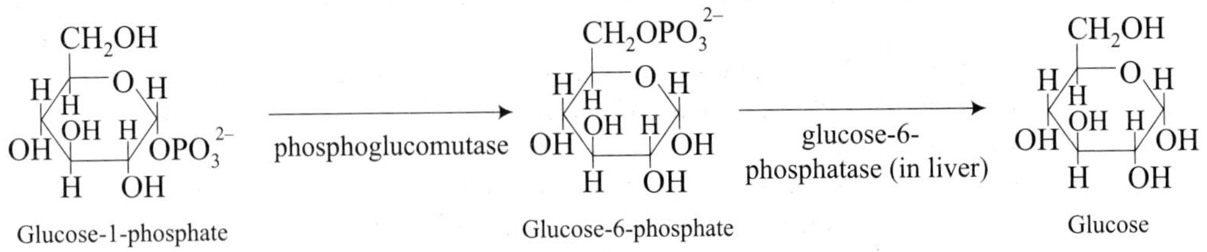

13.6 GLUCONEOGENESIS

Formation of glucose from non-carbohydrate sources is called gluconeogensis. The non-carbohydrate precursors of glucose include lactate, pyruvate, citric acid cycle intermediates and most of the amino acids. Gluconeogenesis mainly occurs in liver and to a lesser extent in kidney. Under the conditions when dietary glucose is not available in sufficient amount, gluconeogensis fulfils the glucose requirement of the body.

13.6.1 Pathway of Gluconeogenesis

Though most of the reactions of gluconeogenesis are the reversal of glycolysis, there are certain reactions which are specific to gluconeogenesis. These are:

  (i) Conversion of pyruvate to phosphoenolpyruvate,

 (ii) Conversion of fructose 1, 6 bisphosphate to fructose 6-phosphate,

(iii) Formation of glucose from glucose-6-phosphate (Fig. 13.10).

  (i) *Conversion of pyruvate to phosphoenolpyruvate*: In the conversion of pyruvate to phosphoenolpyruvate, pyruvate is first converted to oxaloacetate, which is then converted to phosphoenolpyruvate.

Pyruvate is converted to *oxaloacetate* by the enzyme *pyruvate carboxylase* in the presence of ATP and CO_2. Pyruvate carboxylase requires biotin as the coenzyme.

Oxaloacetate is converted to phosphoenolpyruvate by *phosphoenolpyruvate carboxy kinase(PEPCK)*, present in cytosol. High energy phosphate in the form of GTP is required in this reaction.

As oxaloacetate is formed inside the mitochondria it must come out to the cytosol for conversion to phosphoenolpyruvate. As oxaloacetate is not permeable to mitochondria it is converted to malate and cross the mitochondria. In the cytosol malate is again converted to

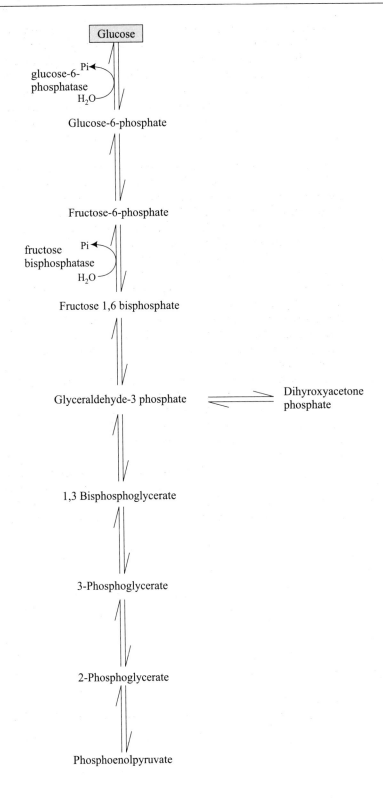

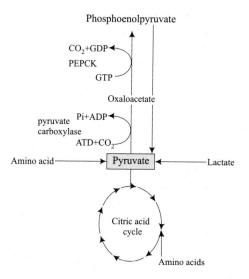

Fig. 13.10 Pathway of gluconeogenesis

oxaloacetate. Oxaloacetate can also combine to acetyl-CoA in the mitochondria to form citrate which is permeable to mitochondria. In the cytosol, citrate is again converted to oxaloacetate.

Once phosphoenolpyruvate is formed it goes into reverse glycolytic pathway to form fructose 1, 6 bisphosphate.

(ii) *Conversion of fructose 1,6 bisphosphate to fructose-6-phosphate*: Fructose 1, 6 bisphosphate is converted to fructose-6-phosphate by a Mg^{2+} dependent enzyme *fructose 1, 6 bisphosphatase*.

$$\text{Fructose 1,6 bisphosphate} \xrightarrow[\text{fructose 1,6 bisphosphatase}]{\overset{H_2O \qquad Pi}{Mg^{2+}}} \text{Fructose-6-bisphosphate}$$

Fructose-6-phosphate is isomerized to glucose-6-phosphate.

(iii) *Formation of glucose from glucose-6-phosphate*: The conversion of glucose-6-phosphate to glucose is a simple hydrolysis of a phosphate ester catalyzed by *glucose-6-phosphatase*. There is no formation of ATP.

$$\text{Glucose-6-phosphate} \xrightarrow[\text{glucose-6-phosphatase}]{H_2O \qquad Pi} \text{Glucose}$$

Glucose produced by gluconeogenesis in liver or kidney is released into blood stream to be carried to the tissues.

Intermediates of citric acid cycle-citrate, isocitrate, α-ketoglutarate, succinyl-CoA, succinate, fumarate and malate enter gluconeogenesis after their oxidation to oxaloacetate. Amino acids enter gluconeogenesis either at the level of pyruvate or after their conversion into an intermediate of citric acid cycle.

13.7 PENTOSE-PHOSPHATE PATHWAY

An alternative pathway for the oxidation of glucose-6-phosphate is pentose phosphate pathway. Glucose-6-phosphate arises either from glycogen breakdown or through the action of hexokinase on glucose. The pathway is also known as *hexose monophosphate shunt* or *phosphogluconate pathway* (Fig. 13.11). Unlike glycolysis and Krebs cycle which lead to the generation of ATP, pentose phosphate pathway leads to the formation of NADPH which is used as an electron donor in many reductive synthesis in the body. The pathway is of particular importance to some tissues especially those involved in lipid biosynthesis such as liver, adipose tissue, mammary gland and adrenal glands.

The pathway consists of two phases:

- Oxidative phase, and
- Non-oxidative phase.

In the oxidative phase, glucose-6-phosphate is converted to a pentose sugar with the formation of NADPH. In the non-oxidative phase, pentose is converted to hexose. During the interconversion of pentose to hexose, three, four and seven carbon sugars are also formed.

13.7.1 Oxidative Phase

(i) *Oxidation of glucose-6-phosphate*: The first step of pentose phosphate pathway is oxidation of glucose-6-phosphate to *6-phosphogluconolactone* in the presence of NADP by the enzyme *glucose-6-phosphate dehydrogenase*. There is net transfer of a hydride ion from C-1 of glucose-6-phosphate to NADP to form NADPH.

Glucose-6-phosphate   6-Phosphogluconolactone

(ii) *Formation of 6-phosphogluconate*: 6-Phosphogluconolactone is hydrolyzed by the enzyme *6-phosphogluconolactonase* to *6-phosphogluconate*.

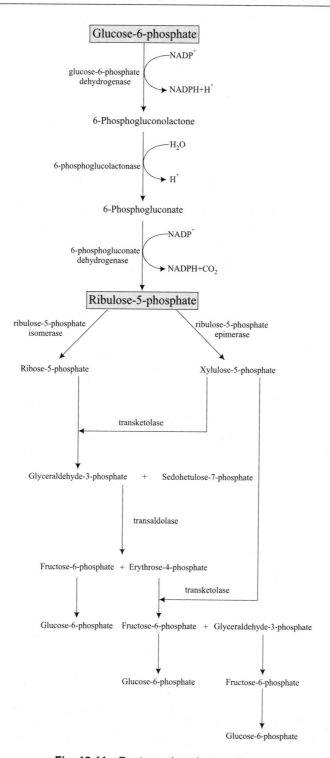

Fig. 13.11 Pentose phosphate pathway

6-Phosphogluconolactone   6-Phosphogluconate

(iii) *Formation of ribulose-5-phosphate*: Enzyme *6-phosphogluconate dehydrogenase* catalyzes the oxidative decarboxylation of 6-phosphogluconate to *ribulose-5-phosphate*. NADP is reduced to NADPH and CO_2 is released.

6-Phosphogluconate   Ribulose-5-phosphate

13.7.2   Non-oxidative Phase

(i) *Isomerization and epimerization of ribulose-5-phosphate*: Ribulose-5-phosphate is acted upon by two different enzymes. *Ribulose-5-phosphate isomerase* converts ribulose-5-phosphate to *ribose-5-phosphate*. *Ribulose-5-phosphate epimerase* converts ribulose-5-phosphate to *xylulose-5-phosphate*.

Amount of ribose-5-phosphate or xylulose-5-phosphate formed, depends on the activity of the cell. Ribose-5-phosphate is relatively high in rapidly dividing cells in which rate of DNA synthesis is high.

(ii) *Conversion of pentose phosphates to hexose-phosphate*: Conversion of pentose phosphates-ribose-5 phosphate and xylulose-5-phosphate to hexose-phosphate (*fructose-6-phosphate*) involves transketolation and transaldolation.

Transketolation: Xylulose-5-phosphate and ribose-5-phosphate react to form *sedoheptulose 7-phosphate* and *glyceraldehyde-3-phosphate*. The reaction is catalyzed by the enzyme *transketolase* which transfers a 2-carbon unit-a ketol group to an aldose.

Ribose-5-phosphate   Xylulose-5-phosphate   Sedoheptulose-7-phosphate   Glyceraldehyde-3-phosphate

Transaldolation: Sedoheptulose-7-phosphate reacts with glyceraldehyde-3-phosphate to form *erythrose-4-phosphate* and *fructose-6-phosphate*. The reaction is catalyzed by transaldolase which transfers a 3-carbon moiety-dihydroxyacetone from ketose to the aldose.

Erythrose-4-phosphate reacts with xylulose-5 phosphate to form *fructose-6-phosphate* and *glyceraldehyde-3-phosphate*. This is another reaction of *transketolation* catalyzed by *enzyme transketolase*.

$$
\begin{array}{ccc}
\underset{\substack{\text{Sedoheptulose-7-phosphate}}}{
\begin{array}{c}
CH_2OH \\
| \\
C=O \\
| \\
HO-C-H \\
| \\
H-C-OH \\
| \\
H-C-OH \\
| \\
H-C-OH \\
| \\
CH_2OPO_3^{2-}
\end{array}}
\ + \
\underset{\substack{\text{Glyceraldehyde-3-phosphate}}}{
\begin{array}{c}
CHO \\
| \\
H-C-OH \\
| \\
CH_2OPO_3^{2-}
\end{array}}
\xrightarrow{\text{transaldolase}}
\underset{\substack{\text{Fructose-6-phosphate}}}{
\begin{array}{c}
CH_2OH \\
| \\
C=O \\
| \\
HO-C-H \\
| \\
H-C-OH \\
| \\
H-C-OH \\
| \\
CH_2OPO_3^{2-}
\end{array}}
\ + \
\underset{\substack{\text{Erythrose-4-phosphate}}}{
\begin{array}{c}
CHO \\
| \\
H-C-OH \\
| \\
H-C-OH \\
| \\
CH_2OPO_3^{2-}
\end{array}}
\end{array}
$$

$$
\begin{array}{ccc}
\underset{\substack{\text{Erythrose-4-phosphate}}}{
\begin{array}{c}
CHO \\
| \\
H-C-OH \\
| \\
H-C-OH \\
| \\
CH_2OPO_3^{2-}
\end{array}}
\ + \
\underset{\substack{\text{Xylulose-5-phosphate}}}{
\begin{array}{c}
CH_2OH \\
| \\
C=O \\
| \\
HO-C-H \\
| \\
H-C-OH \\
| \\
CH_2OPO_3^{2-}
\end{array}}
\xrightarrow{\text{transketolase}}
\underset{\substack{\text{Fructose-6-phosphate}}}{
\begin{array}{c}
CH_2OH \\
| \\
C=O \\
| \\
HO-C-H \\
| \\
H-C-OH \\
| \\
H-C-OH \\
| \\
CH_2OPO_3^{2-}
\end{array}}
\ + \
\underset{\substack{\text{Glyceraldehyde-3-phosphate}}}{
\begin{array}{c}
CHO \\
| \\
H-C-OH \\
| \\
CH_2OPO_3^{2-}
\end{array}}
\end{array}
$$

Fructose-6-phosphate is isomerized to form glucose-6-phospate. Glyceraldehyde-3-phosphate is also converted to glucose-6-phosphate by the enzymes of glycolytic pathway working in reverse direction. With the formation of glucose-6-phosphate, the whole cycle is repeated again.

13.7.3 Significance of Pentose Phosphate Pathway

Pentose phosphate pathway leads to the formation of NADPH. NADPH is used as an electron donor for the synthesis of fatty acids, steroids, cholesterol and sphingolipids. Tissues most heavily involved in lipid biosynthesis are rich in pentose phosphate pathway enzymes.

NADPH is also used to counter the damaging effects of oxygen radicals in the cells directly exposed to free radicals such as erythrocytes and cells of lens and cornea. Pentose sugar, formed during pentose phosphate pathway is used by the cells to make nucleic acids and coenzymes. CO_2 produced during pentose phosphate pathway is used in CO_2 fixation by plants.

13.8 REGULATION OF BLOOD GLUCOSE

In a normal healthy individual the blood glucose ranges from 80 – 120 mg/100 ml of blood in fasting state and 100 – 140 mg/100 ml of blood after a meal. The concentration of glucose in the blood depends on a balance between the amount of glucose that enters the blood and the amount of glucose which is removed from the blood.

Glucose enters blood stream through diet, gluconeogenesis and glycogenolysis. Glucose is removed from the blood by glycogenesis, lipogenesis, excertion in urine and oxidation of glucose by the tissues.

The rise in blood sugar above the normal range is called as *hyperglycemia*. When the blood sugar falls the normal range, the condition is termed as *hypoglycemia*. There are various factors which maintain the blood glucose level within the normal range.

13.8.1   Regulation during Hyperglycemia

After a carbohydrate rich diet, blood glucose level begins to rise. To maintain the blood glucose to its normal level, insulin is secreted by the β-cells of islets of langerhans of pancreas, soon after a meal. Insulin is a hypoglycemic hormone – which lowers the blood glucose level. The action of insulin is at different levels. Insulin enhances the utilization of glucose in the peripheral tissues by increasing the permeability of the cell membrane for glucose. Insulin has a stimulatory effect on glycogen synthesis both in liver and muscles. It also suppresses gluconeogenesis. In liver and adipose tissues insulin helps in the synthesis of fatty acids. These actions of insulin result in the decrease of blood glucose level and help to maintain it at its normal range.

13.8.2   Regulation during Hypoglycemia

Under the condition of hypoglycemia, a number of factors contribute to increase the blood glucose level. Several hours after a meal blood glucose level begins to drop. This leads to a decrease in insulin secretion and increase in glucagon secretion. Glucagon, a hyperglycemic hormone, is produced by the α-cells of islets of langerhans of pancreas. It increases blood glucose level by its different actions. Glucagon mobilizes glycogen stores and stimulates degradation of liver glycogen. It inhibits glycogen synthesis. The effect is by triggering the cyclic AMP cascade which leads to the phosphorylation and activation of phosphorylase and the inhibition of glycogen synthase. Glucagon also stimulates gluconeogenesis in liver. It inhibits fatty acid synthesis by decreasing the production of pyruvate and by lowering the activity of acetyl-CoA. These different actions of glucagon increase the blood glucose level.

Besides glucagon, glucocorticoids also have stimulatory effect on gluconeogenesis in liver. Glucocorticoids also cause the mobilization of amino acids from extrahepatic tissues mainly from muscles.

Other hormones such as GH and thyroxine also act as hyperglycemic hormones and help in maintaining the normal blood glucose level in the body.

13.9   SUMMARY

- Carbohydrates are metabolized by different pathways to produce energy.
- Glycolysis is the sequence of reactions which result in the formation of two molecules of pyruvate from one glucose molecule with the net gain of two ATP and two NADH.
- Pyruvate undergoes different metabolic pathways under different conditons.
- Under anaerobic conditions pyruvate forms lactic acid.

- In alcoholic fermentation, pyruvate is decarboxylated to yield ethanol and CO_2.
- Under aerobic conditions pyruvate forms acety-CoA which is further metabolized by citric acid cycle.
- Citric acid cycle involves eight enzymatic reactions and results in the oxidation of acetyl group to CO_2 with net gain of three NADH, one $FADH_2$ and one GTP per cycle.
- The formation of glycogen from glucose is called glycogenesis. In glycogenesis, glucose-1-phosphate is activated in presence of UTP to form UDP-glucose. UDP-glucose is converted to glycogen by the enzyme glycogen synthase and debranching enzymes.
- The breakdown of glycogen is called glycogenolsis. Glycogen phosphorylase converts the glucosyl units at the non-reducing end of glycogen to glucose-1-phosphate. Glucose-1-phosphate is converted to glucose-6-phosphate which is hydrolyzed to glucose by glucose-6-phosphatase.
- Gluconeogenesis is the formation of glucose from non-carbohydrate sources. Most of the reactions of gluconeogenesis are glycolytic reactions that proceed in reverse.
- Pentose phosphate pathway is an alternative pathway for the oxidation of glucose. In this pathway, glucose-6-phospahte is oxidized and decarboxylated to produce NADPH, CO_2 and pentose sugar.
- Blood glucose-level remains constant under normal conditions. There are various factors which regulate the blood glucose level in the body.

EXERCISE

1. What is glycolysis? Summarize various steps in glycolysis.
2. Outline the reactions of citric acid cycle.
3. Describe pentose phosphate pathway. Write its significance.
4. Write notes on:
   (i) Regulation of blood glucose level.
   (ii) Fate of pyruvate
   (iii) Glycogenolysis

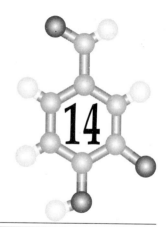

Metabolism of Lipids

Fats are the major constituents of dietary lipids. Digestion of fats mainly occurs in the intestine where enzyme pancreatic lipase breaks down fats into fatty acids and glycerol. Digestion of fats also occurs in the mouth and stomach in very small amount. But the major site of fat digestion is the small intestine. Fats are water insoluble compounds and enzyme lipases are water soluble.

In the intestine, first bile salts break the large fat particles into smaller fine particles. The process is known as *emulsification*. Emulsification increases the surface area of the fats for the action of enzyme lipase. Digestion of fats takes place at lipid-water interface. Enzyme lipase breaks down fats (triacylglycerol) into fatty acids-1, 2 diacylglycerol and 1, acylglycerol.

14.1 ABSORPTION AND TRANSPORT

Fatty acids and di-and monoacylglycerols are absorbed by the intestinal mucosa. Bile salts also help in the absorption. After their absorption, these products of digestion-fatty acids, di-and monoacylglycerols get resynthesized into triacylglycerols. These triacylglycerols are packaged in the form of lipoprotein particles known as *chylomicrons*. These chylomicrons are passed through the circulatory system to other tissues. On reaching the specific tissues i.e. liver and adipose tissues, the triacylglycerols of chylomicrons are hydrolyzed by the enzyme lipoprotein lipase into monoacylglycerol and fatty acids which are absorbed and stored in the tissues in the form of triacylglycerols.

14.2 OXIDATION OF FATTY ACIDS

The degradation of fat in the body begins with the breakdown of neutral fat into glycerol and fatty acids by the enzyme lipase. Glycerol reacts with ATP to form glycerol phosphate which is oxidized to form glyceraldehyde-3-phosphate. Glyceraldehyde-3-phosphate may either form pyruvate and enter TCA cycle or it may also be utilized to form glucose or glycogen.

Fatty acids are subjected to oxidation. Principal method of fatty acids oxidation is *β-oxidation*. β-oxidation of fatty acids was proposed by *Franz Knoop* in 1904. On the basis of his experiments Knoop deduced that fatty acids are progressively degraded by two carbon units by the oxidation of β-carbon atom.

The enzymes for oxidation of fatty acids are located in the mitochondrial matrix in animal cells. Thus, fatty acids are to enter the mitochondria. Before the fatty acids enter the mitochondria, they are first activated.

Activation of fatty acids: The activation of fatty acids is catalyzed by a family of *acetyl-CoA synthetases* isozymes, present in the outer mitochondrial membrane. Different isozymes differ in their chain-length specificities. Acetyl-CoA synthetases catalyze the formation of a thioester linkage between the fatty acid carboxyl group and the thiol group of coenzyme-A which results in the formation of *fatty acyl-CoA*.

$$\text{Fatty acid} + \text{Co A} + \text{ATP} \rightleftharpoons \text{fatty acyl-CoA} + \text{AMP} + \text{PPi}$$

Fatty acyl-CoA are high energy compounds. They are then transported to mitochondria for oxidation.

Transport of fatty acids: The short chain fatty acids (fatty acids with chain length of 12 or fewer carbon) enter mitochondria whithout the help of membrane transporters. But a long chain fatty acyl-CoA cannot directly cross the inner mitochondrial membrane. Such fatty acids are transiently attached to *carnitine* to form *fattyacyl carnitine*. The reaction is catalyzed by enzyme *carnitine acyl transferase I*, present in the outer membrane of mitochondria. Fatty acyl-carnitine then enters the matrix. A specific carrier protein transports acyl-carnitine into the mitochondria. Now, the fatty acyl group is transferred from carnitine to intra mitochondrial coenzyme A to form fatty acyl-CoA. This reaction is catalyzed by the enzyme *carnitine acyl transferase II*, located on the inner surface of the inner membrane. Free carnitine is transported back by the same carrier protein which transports acyl-carnitine into the mitochondria (Fig. 14.1). Once inside the mitochondria, fatty acyl-CoA is oxidized in four enzymatic reactions as follows:

14.2.1 β-oxidation

The first reaction in the β-oxidation of fatty acyl-Co A is the formation of *trans-Δ^2-enoyl-CoA* by the flavo enzyme *acyl-CoA dehydrogenase*. There are three isozymes of acyl-CoA dehydrogenase which are specific for short, medium and long chain fatty acid-CoAs. The electrons removed from fatty acyl-CoA are transferred to FAD to form $FADH_2$. $FADH_2$ is reoxidized by mitochondrial electron transport chain through a series of electron transfer reactions.

In the second reaction, there is hydration of the double bond of trans-Δ^2-enoyl-Co A to form *β-hydroxyacyl-CoA*. The reaction is catalyzed by the enzyme *enoyl-CoA hydratase*. In the third reaction, β-hydroxy acyl-CoA is dehydrogenated to form *β-ketoacyl-CoA*. The reaction is catalyzed by the enzyme *β-hydroxyacyl-CoA dehydrogenase*. NAD^+ is the electron acceptor in this reaction.

In the final reaction of β-oxidation, there is cleavage of $C_\alpha - C_\beta$ of β-ketoacyl-CoA to form *acetyl-CoA* and a new *acyl-CoA*, which is two carbon atom shorter than the original one. This reaction is catalyzed by *β-ketoacyl-CoA thiolase* (or *thiolase*). Acetyl-CoA enters the citric acid cycle and is oxidized to CO_2 and H_2O (Fig. 14.2).

14.2.2 Energetics of β-oxidation

During each round of β-oxidation one molecule of acetyl-CoA, one NADH and one $FADH_2$ are produced. Further oxidation of acetyl-CoA by citric acid cycle yields 3 NADH, 1 $FADH_2$ and 1 GTP

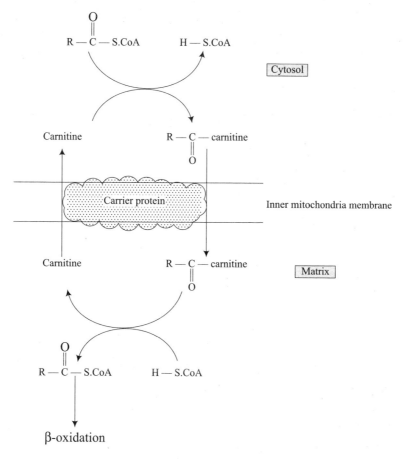

Fig. 14.1 Transport of fatty acids

molecules. NADH is reoxidized to yield 3ATP. One $FADH_2$ produces 2 ATP through oxidative phosphorylation. In the oxidation of palmitic acid $(C_{15}H_{31}COOH)$, there are 7 rounds of β-oxidation which yield 8 molecules of acetyl-CoA, 7 NADH and 7 $FADH_2$. These acetyl-CoA produce 24 NADH, 8$FADH_2$ and 8GTP by citric acid cycle. Oxidative phosphorylation of these 31 NADH and 15 $FADH_2$ produce 93 and 30 ATP respectively. Thus, total 131 ATP are produced by complete oxidation of palmitic acid. As 2ATP are utilized in the activation of fatty acid, the net gain is 129 ATP. This can be summerized as:

- Total molecules of acetyl-CoA produced in seven rounds   = 8
- ATP produced in each round of β-oxidation   = 5
- ATP produced in seven rounds $= 7 \times 5 = 35$
- ATP produced by oxidation of acetyl-CoA by citric acid cycle (12×8) = 96

  Total ATP produced   = 131

  ATP utilized in activation of fatty acid   = (–) 02

  Net gain   = 129

$$CH_3 - (CH_2)_n - \overset{\overset{\displaystyle H}{|}\beta}{\underset{\underset{\displaystyle H}{|}}{C}} - \overset{\overset{\displaystyle H}{|}\alpha}{\underset{\underset{\displaystyle H}{|}}{C}} - \overset{\overset{\displaystyle O}{\|}}{C} - OH \qquad \text{Free fatty acid}$$

Cytosol

CoA-SH $\diagdown$ $\diagup$ ATP

acyl-CoA sythetase $\longrightarrow$ AMP+PPi

$$CH_3 - (CH_2)_n - \overset{\overset{\displaystyle H}{|}}{\underset{\underset{\displaystyle H}{|}}{C}} - \overset{\overset{\displaystyle H}{|}}{\underset{\underset{\displaystyle H}{|}}{C}} - \overset{\overset{\displaystyle O}{\|}}{C} - S.CoA \qquad \text{Fatty acyl CoA.}$$

- - - - - - - - - - Carnitine - - - - - - - - - -

Mitochondrial Membrane

$$CH_3 - (CH_2)_n - \overset{\overset{\displaystyle H}{|}\beta}{\underset{\underset{\displaystyle H}{|}}{C}} - \overset{\overset{\displaystyle H}{|}\alpha}{\underset{\underset{\displaystyle H}{|}}{C}} - \overset{\overset{\displaystyle O}{\|}}{C} - S.CoA$$

acyl-CoA dehydrogenase $\diagup$ FAD

$\longrightarrow$ FADH$_2$

$$CH_3 - (CH_2)_n - \overset{\overset{\displaystyle H}{|}}{C} = \overset{\overset{\displaystyle H}{|}}{\underset{\underset{\displaystyle H}{|}}{C}} - \overset{\overset{\displaystyle O}{\|}}{C} - S.CoA \qquad \text{trans-}\Delta^2\text{-Enoyl-CoA}$$

enoyl-CoA hydratase $\diagup$ H$_2$O

$$CH_3 - (CH_2)_n - \overset{\overset{\displaystyle H}{|}}{\underset{\underset{\displaystyle OH}{|}}{C}} - CH_2 - \overset{\overset{\displaystyle O}{\|}}{C} - S.CoA \qquad \beta\text{-Hydroxyacyl-CoA}$$

β-hydroxyacyl CoA dehydrogenase $\diagup$ NAD+

$\longrightarrow$ NADH+H$^+$

$$CH_3 - (CH_2)_n - \overset{\overset{\displaystyle O}{\|}}{C} - CH_2 - \overset{\overset{\displaystyle O}{\|}}{C} - S.CoA \qquad \beta\text{-Ketoacyl-CoA}$$

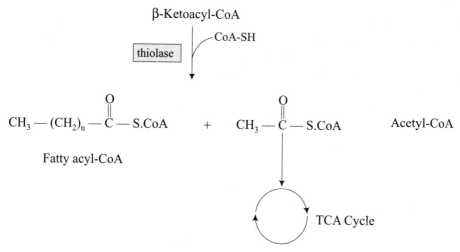

Fig. 14.2 β-oxidation of fatty acids

14.2.3   Peroxisomal β-oxidation

Besides mitochondria, some β-oxidation occurs in peroxisomes. Fatty acid oxidation in peroxisomes shortens very long chain of fatty acids which are then fully degraded by β-oxidation in mitochondria. In case of plants, fatty acid oxidation occurs exclusively in peroxisomes and glyoxysomes. In peroxisomal β-oxidation same chemical changes occur in fatty acids as in the mitochondria. The difference is in the initial dehydrogenation reaction. In this reaction, electrons are transferred directly to O_2 to form H_2O_2 instead of formation of $FADH_2$ as in the case of mitochondrial β-oxidation. H_2O_2 is converted to H_2O and O_2 by the enzyme *catalase*.

Further reactions are similar to mitochondrial β-oxidation reactions. Since $FADH_2$ is not formed in peroxisomal β-oxidation, energy production (ATP formation) per cycle by peroxisomal β-oxidation is less in comparison to mitochondrial β-oxidation.

14.2.4 ω-oxidation of the Fatty Acids

In some animals, oxidation of fatty acids occurs at ω-(omega) carbon atom i.e. the carbon atom situated farthest from the carboxyl group.

$$
\underset{\omega}{CH_3}-CH_2-CH_2-(CH_2)_n-\underset{\beta}{CH_2}-\underset{\alpha}{CH_2}-COOH \qquad \text{(Fatty acid)}
$$

ω-oxidation

$$
COOH-CH_2-CH_2-(CH_2)_n-CH_2-CH_2-COOH \qquad \text{(Dicarboxylic acid)}
$$

β-oxidation

$$
COOH-(CH_2)_n-COOH \qquad \text{(Smaller dicarboxylic acid)}
$$

ω-oxidation results in the production of dicarboxylic acid. Dicarboxylic acid undergoes β-oxidation and forms successively smaller dicarboxylic acids.

14.2.5 α-oxidation of Fatty Acids

This is another alternative pathway of oxidation of fatty acids. In this pathway, oxidation takes place by the removal of one carbon atom at a time from the carboxyl end of the fatty acid. Fatty acid is, first, decarboxylated and reduced to an aldehyde with one carbon less than the original fatty acid. The aldehyde is then oxidized to form the acid which subsequently undergoes repeated β-oxidation.

$$
R-CH_2-CH_2-CH_2-COOH
$$

peroxidase $\quad$ $-H_2O_2$

$\rightarrow H_2O + CO_2 + O_2$

$$
R-CH_2-CH_2-CHO \qquad \text{(Aldehyde)}
$$

dehydrogenase $\quad$ $-NAD+$

$\rightarrow NADH + H^+$

$$
R-CH_2-CH_2-COOH \qquad \text{(Acid)}
$$

14.2.6   Ketone Bodies

The oxidation of fatty acids produces acetyl-CoA which undergoes citric acid cycle. Under certain metabolic conditions associated with high rate of oxidation of fatty acids, in liver cells, acetyl-CoA is converted to *acetoacetate* and *β-hydroxybutyrate*. Acetoacetate continually undergoes spontaneous decarboxylation to produce *acetone*. These three compounds i.e. acetoacetate, β-hydroxybutyrate and acetone are collectively called as *ketone bodies*. The concentration of ketone bodies in the blood of a well-fed individual does not normally exceed 1 mg/100 ml of blood. When the concentration of ketone bodies in the blood rises above the normal level, the condition is known as *ketonaemia*. When the blood level of ketone bodies rises above the renal threshold, they are excreted in urine. This condition is *ketonuria*. The term *ketosis* refers to the accumulation of abnormal amounts of ketone bodies in tissues and body fluids.

Ketogenesis: The first step in the formation of ketone bodies (ketogenesis) is the condensation of two molecules of acetyl-Co A by the enzyme *thiolase* to produce *acetoacetyl-CoA*. The reaction is the reversal of the last step of β-oxidation. Acetoacetyl-CoA, then, condenses with one another molecule of acetyl-CoA form *β-hydroxy-β-methylglutaryl-CoA (HMG-CoA)*. The reaction is catalyzed by *HMG-CoA synthase*. HMG-CoA is then cleaved to *acetoacetate* and *acetyl-CoA* by the enzyme *HMG-CoA lyase*. Acetoacetate may be reduced to *β-hydroxy butyrate* by the enzyme *β-hydroxy butyrate dehydrogenase*. Acetoacetate also undergoes decarboxylation to form *acetone* and CO_2. (Fig. 3)

Significance of ketone bodies: After their formation, liver releases the ketone bodies into the blood which carries them to extra-hepatic tissues. Ketone bodies are an important metabolic fuel for many peripheral tissues particularly heart and skeletal muscles. During starvation, brain also uses ketone bodies as the major fuel source. The production and export of ketone bodies by the liver also allows the continued oxidation of fatty acids with minimal oxidation of acetyl-CoA.

14.3   FATTY ACID SYNTHESIS

Fatty acid biosynthesis takes place in the cytosol of the cell. The starting material for fatty acid synthesis is acetyl-CoA. Acetyl-CoA is synthesized in the mitochondria by the oxidative phosporylation of pyruvate. Acetyl-CoA enters the cytosol in the form of citrate.

The first step in the synthesis of fatty acids is the carboxylation of acetyl-CoA to *malonyl-CoA*. The reaction is catalyzed by the enzyme *acetyl-CoA carboxylase* which is a biotin dependent enzyme. Mn^{2+} is required as a cofactor and ATP provides the energy (Fig. 14.4).

Once malonyl-CoA is formed, the molecule is acylated several times to produce fatty acid-palmitic acid. These reactions are catalyzed by a multienzyme complex-*fatty acid synthase*.

Fatty acid synthase: In yeast, mammals and birds fatty acid synthase is a multienzyme complex which is made up of an ellipsoid dimer of the two identical polypeptide monomeric units arranged in a head to tail fashion. Each of these units consists of six enzymes and an acyl carrier protein (ACP) molecule. The complex is functional only when the two monomeric units are in association with each other. In E. coli, the individual enzymes are separate.

During the subsequent reactions of fatty acid synthesis growing fatty acid remains attached at acyl carrier protein (ACP) in the enzyme.

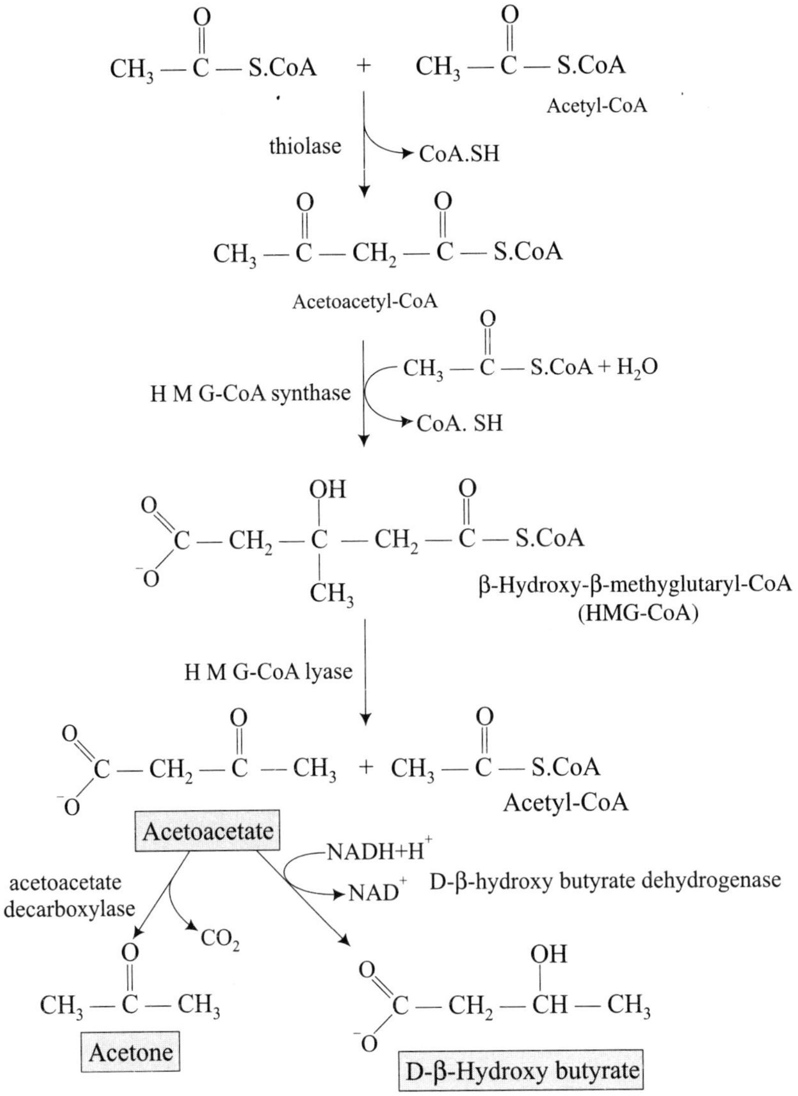

$$CH_3 - \overset{\overset{\displaystyle O}{\|}}{C} - S.CoA \quad + \quad CH_3 - \overset{\overset{\displaystyle O}{\|}}{C} - S.CoA$$

Acetyl-CoA

thiolase $\longrightarrow$ CoA.SH

$$CH_3 - \overset{\overset{\displaystyle O}{\|}}{C} - CH_2 - \overset{\overset{\displaystyle O}{\|}}{C} - S.CoA$$

Acetoacetyl-CoA

H M G-CoA synthase
$$CH_3 - \overset{\overset{\displaystyle O}{\|}}{C} - S.CoA + H_2O$$
$\longrightarrow$ CoA. SH

$$\overset{\overset{\displaystyle O}{\diagdown}}{\underset{\displaystyle O}{\diagup}} C - CH_2 - \overset{\overset{\displaystyle OH}{|}}{\underset{\displaystyle CH_3}{C}} - CH_2 - \overset{\overset{\displaystyle O}{\|}}{C} - S.CoA$$

β-Hydroxy-β-methyglutaryl-CoA
(HMG-CoA)

H M G-CoA lyase

$$\overset{\overset{\displaystyle O}{\diagdown}}{\underset{\displaystyle O}{\diagup}} C - CH_2 - \overset{\overset{\displaystyle O}{\|}}{C} - CH_3 \quad + \quad CH_3 - \overset{\overset{\displaystyle O}{\|}}{C} - S.CoA$$

Acetyl-CoA

Acetoacetate

NADH+H$^+$

acetoacetate decarboxylase

NAD$^+$ D-β-hydroxy butyrate dehydrogenase

CO$_2$

$$CH_3 - \overset{\overset{\displaystyle O}{\|}}{C} - CH_3$$

Acetone

$$\overset{\overset{\displaystyle O}{\diagdown}}{\underset{\displaystyle O}{\diagup}} C - CH_2 - \overset{\overset{\displaystyle OH}{|}}{CH} - CH_3$$

D-β-Hydroxy butyrate

Fig. 14.3 Ketogenesis

Acetyl-CoA combines with ACP molecule of one of the monomeric chains to form *acetyl-ACP*. In the process, CoA is removed and acetyl group is transferred to ACP. Similarly, a malonyl group is transferred from malonyl-CoA to ACP of other monomeric unit to form *malonyl-ACP*. These reactions are catalyzed by enzyme *transacylase*.

In the condensation reaction, the malonyl-ACP is decarboxylated and attacks the acetyl-thioester to form *acetoacetyl-ACP*. The reaction is catalyzed by the enzyme *β-ketoacyl-ACP synthase*.

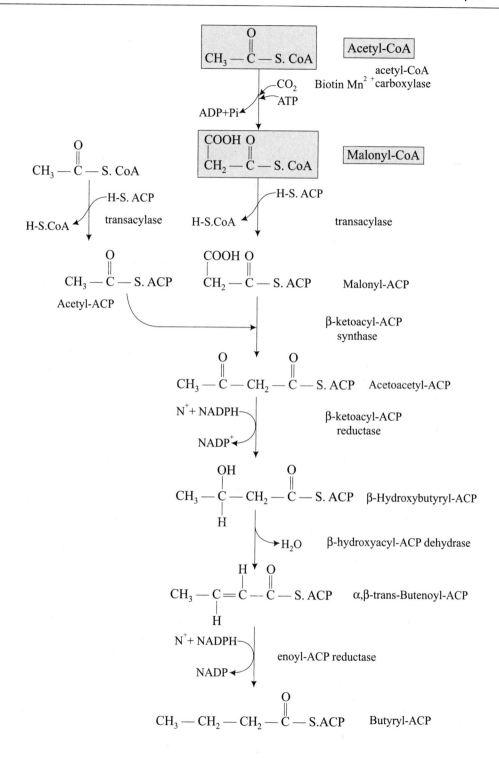

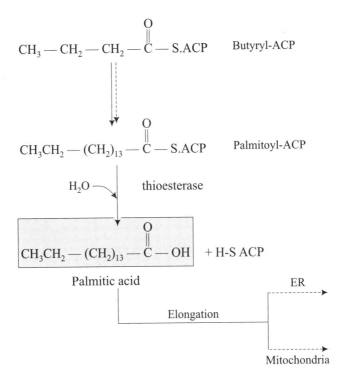

Fig. 14.4 Biosynthesis of fatty acids

Ketoacyl group of acetoacetyl-ACP is reduced to hydroxy group to form *β-hydroxy-butyryl ACP*. The reaction is catalyzed by the enzyme *β-ketoacyl-ACP reductase*.

β-hydroxy butyryl-ACP undergoes dehydration by the enzyme *β-hydroxyacyl-ACP dehydrase* to form *α, β-trans-butenoyl-ACP* which is reduced to form *butyryl-ACP*. This reduction is catalyzed by the enzyme *enoyl-ACP reductase*, using NADPH as coenzyme.

Once butyryl-ACP is formed, these reactions are repeated six more times to form *palmitoyl-ACP*. Palmitoyl-ACP is released from the enzyme complex as *palmitic acid* by the enzyme *thioesterase*.

14.3.1 Elongation of Fatty Acids

Once palmitic acid is formed, it is converted to longer chain saturated fatty acids by the enzymes *elongases*, present in mitochondria and endoplasmic reticulum. In endoplasmic reticulum, successive condensation of malonyl-CoA with acetyl-CoA results in the formation of longer chain fatty acids. In case of mitochondria, acetyl-CoA is directly incorporated into the palmityl-CoA. Neither ACP nor malonyl-CoA is used here.

14.4 BIOSYNTHESIS OF CHOLESTEROL

Cholesterol is an important molecule in animals. Biosynthesis of cholesterol is a complex process. Essentially all tissues in the body form cholesterol. Liver is the major site of cholesterol biosynthesis. Besides liver, adrenal cortex, gonads, skin and intestine are the other organs which are actively involved in the biosynthesis of cholesterol.

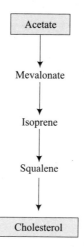

Cholesterol

Cholesterol biosynthesis was first elucidated by Konrad Bloch. All carbon atoms of cholesterol are derived from acetate. In the biosynthesis of cholesterol first acetate condenses to form *mevalonate* which is converted to *isoprene units*. Isoprene units then condense to form the 30-carbon linear *squalene*. Cyclization of squalene results in the formation of four ring structure of cholesterol (Fig. 14.5).

```
        ┌─────────────┐
        │   Acetate   │
        └─────────────┘
               │
               ▼
          Mevalonate
               │
               ▼
           Isoprene
               │
               ▼
           Squalene
               │
               ▼
        ┌─────────────┐
        │ Cholesterol │
        └─────────────┘
```

Fig. 14.5 Biosynthesis of cholesterol

Synthesis of Mevalonate: The first step in the synthesis of cholesterol is the formation of *β-hydroxy-β-methylglutaryl-CoA (HMG-CoA)*. Two molecules of acetyl-CoA condense to form acetoacetyl-CoA which condenses with one another molecule of acetyl-CoA to produce HMG-CoA. The first reaction is catalyzed by enzyme *thiolase*. The conversion of acetoacetyl-CoA to HMG-CoA requires *HMG-CoA synthase*. Cytosolic isozymes of these two enzymes i.e. thiolase and HMG-CoA synthase are distinct from mitochondrial isozymes that are involved in ketone body synthesis. HMG-CoA is then reduced to form *mevalonate* by the enzyme *HMG-CoA reductase*. This is the

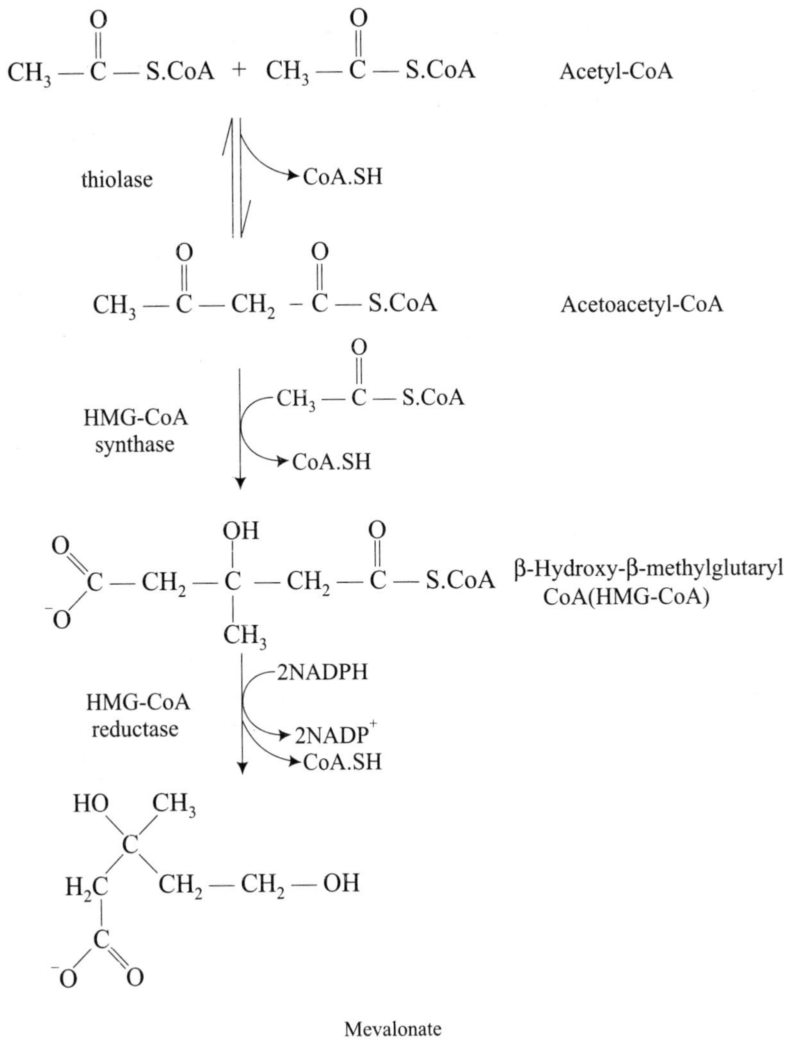

Fig. 14.6 Synthesis of mevalonate

rate-limiting step in the biosynthesis of cholesterol. Cholesterol itself inhibits the enzyme HMG-CoA reductase by 'feed-back-inhibition' (Fig. 14.6).

Glucagon and glucocorticoids also decrease the activity of this enzyme. Insulin has a stimulatory effect on HMG-CoA reductase activity.

Formation of Isoprene Units: Mevalonate is phosphorylated by ATP to form several phosphorylated intermediates. *Isopentenyl pyrophosphate* is the first active isoprene unit. Isopentenyl undergoes isomerization to from second activated isoprene *dimethylallyl pyrophosphate* (Fig. 14.7).

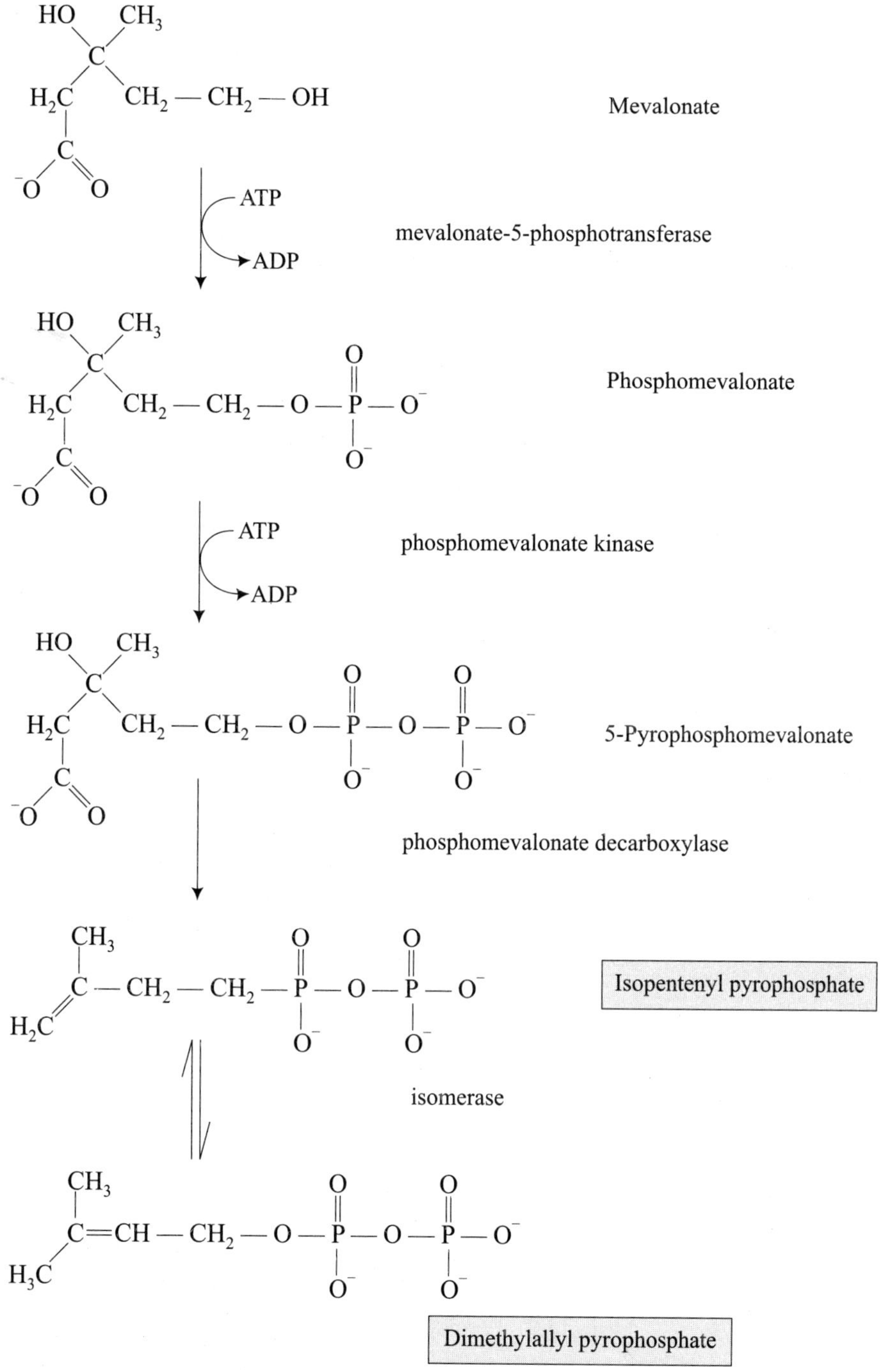

Fig. 14.7 Formation of isoprene units

Formation of Squalene: Isopentenyl pyrophospate and dimethylallyl pyrophosphate condense to form *geranyl pyrophosphate* in the presence of enzyme *prenyl transferase*. Geranyl pyrophosphate condenses with another molecule of isopentenyl pyrophosphate to form *farnesyl pyrophosphate*. Finally, two molecules of farnesyl pyrophosphate condense to form squalene with the elimination of both pyrophosphate groups (Fig. 14.8).

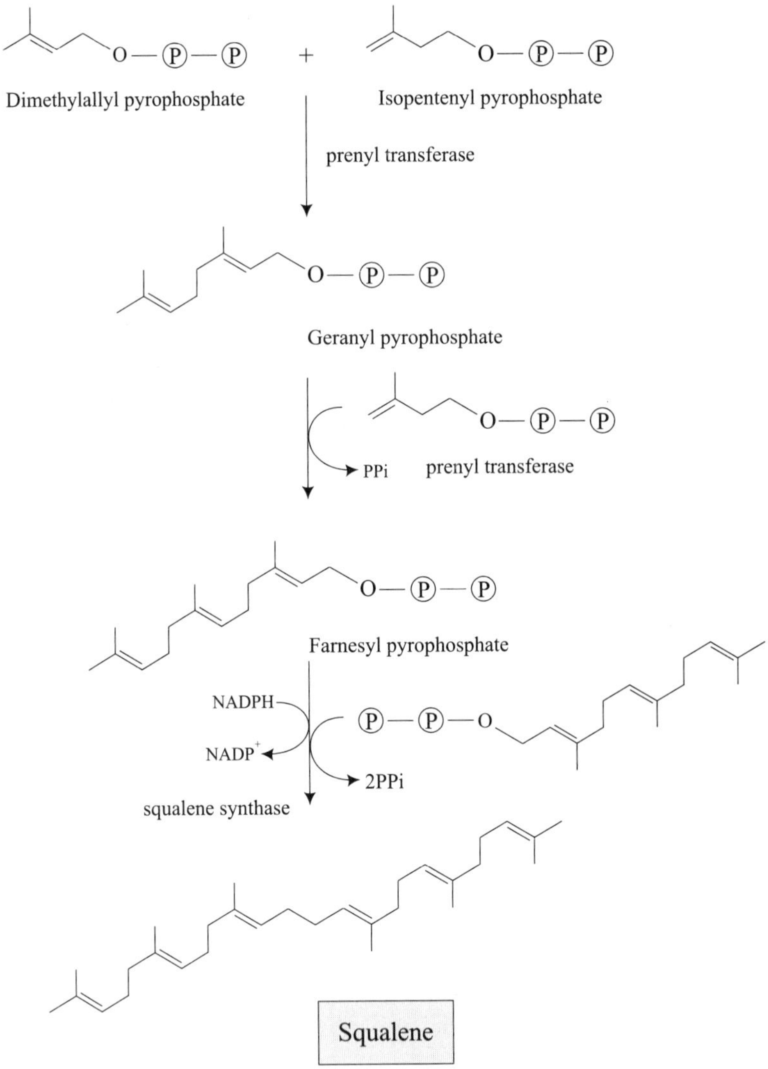

Fig. 14.8 Formation of squalene

Formation of Cholesterol: Squalene is oxidized to form *squalene-2, 3 epoxide* by the enzyme *squalene monooxygenase*. Enzyme *cyclase* brings about the cyclization of squalene- 2, 3, epoxide to form *lanosterol*. Finally, lanosterol is converted to cholesterol by a series of reactions (Fig. 14.9).

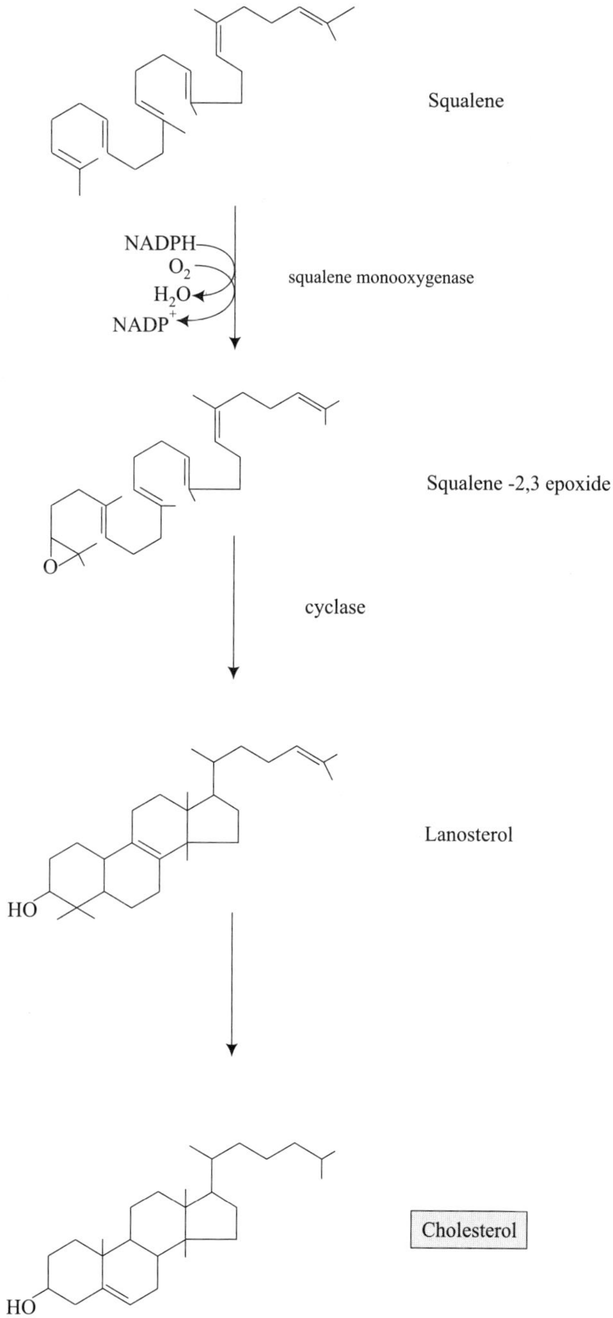

Fig. 14.9   Formation of cholesterol from squalene

14.4.1 Fate of Cholesterol

There are two sources of cholesterol in the body-it is synthesized in various organs or it can be obtained from diet.

Cholesterol synthesized by the liver is either converted to bile acids or esterified by the enzyme *acyl-CoA cholesterol acyltransferase (ACAT)* to form cholesteryl esters. Cholesteryl esters are transported to other tissues which use cholesterol. Cholesteol is transported in the blood in the form of lipoprotein particles. A lipoprotein particle consists of a core of hydrophobic lipids surrounded by a shell of more polar lipids and proteins. There are several types of lipoprotein particles such as chylomicrons, chylomicron remnants, very low density lipoproteins (VLDL), intermediate density lipoproteins (IDL), low density lipoproteins (LDL) and high density lipoproteins (HDL). Cholesterol synthesized in the body, is transported in the form of low density lipoproteins (LDL). Cholesteryl esters are first packaged in the form of VLDL which is then converted to IDL and ultimately form LDL. LDL are the major carrier of cholesterol in blood. LDL transport cholesterol to the tissues. Inside the tissue cholesteryl esters are hydrolyzed to free cholesterol. Cholesterol has several fates in the body (Fig. 14.10).

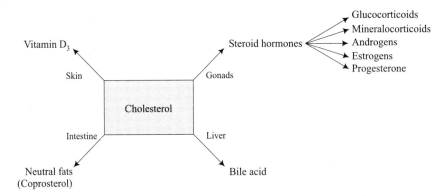

Fig. 14.10 Fate of cholesterol

- More then 50% of cholesterol is converted to bile acids.
- About 10% of cholesterol is converted to neutral sterols called coprosterol.
- Cholesterol is a precursor of vitamin D. In skin, cholesterol (7-dehydrocholesterol) is converted to cholecalciferol (vitamin D_3) by UV rays.
- Cholesterol is used as a precursor for the synthesis of several steroid hormones-adrenocorticoids, androgens, estrogens and progesterone.

14.5 SUMMARY

- Fats are the major form of metabolic energy storage in our body.
- Enzyme lipase breaks fats into fatty acids and glycerol. Bile salts help in the process through their emulsifying activity.

- β-oxidation is the main pathway of fatty acid metabolism. Fatty acids are first activated by the enzyme acetyl-CoA synthetase.

- β-oxidation occurs in four enzymatic reactions—(i) formation of trans-Δ^2-enoyl CoA. (ii) hydration of the double bond of trans-Δ^2-enoyl CoA. (iii) dehydration of β-hydroxy acetyl-CoA to form β-ketoacyl-CoA, and (iv) cleavage of α,β carbon atoms of β-ketoacyl-CoA to form acetyl-CoA.

- There is net gain of one molecule of acetyl-CoA, one molecule of NADH and one molecule of $FADH_2$ per cycle of β-oxidation.

- ω-oxidation and α-oxidation are the alternative pathways of oxidation of fatty acids.

- In the liver, under certain metabolic conditions, acetyl-CoA is converted into acetoacetate, β-hydroxybutyrate and acetone. These are called ketone bodies. Ketone bodies serve as metabolic fuel for some body tissues.

- Fatty acids are synthesized by acetyl-CoA. Acetyl-CoA is converted to malonyl-CoA which is further acylated several times to produce palmitic acid. Longer chain fatty acids are synthesized from palmitic acid through the action of enzymes elongases.

- Cholesterol biosynthesis is a lengthy process. Acetyl-CoA first forms mevalonate which is converted to isoprene units. Six isoprene units condense to form squalene. Finally, squalene cyclizes to form cholesterol. Cholesterol has several fates in the body.

EXERCISE

1. Summarize the reactions of β-oxidation of fatty acid.
2. How are ketone body synthesized in the body? Write their significance.
3. Describe the pathway of cholesterol biosynthesis.
4. Write short notes on:
    (i) Fate of cholesterol,
    (ii) Biosynthesis of fatty acids

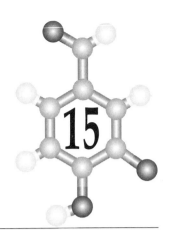

Metabolism of Proteins

Proteins are the building materials in the body. Different types of proteins perform different functions. They are made up of amino acids. During the process of digestion, proteins are hydrolyzed into amino acids as an end product. After absorption in the small intestine, amino acids are carried to the liver by hepatic portal system. From here, they enter the systemic circulation and diffuse throughout the body fluids. In the body, amino acids are also formed by the hydrolysis of protoplasm of worn out tissues. At the same time, there is a continuous synthesis of amino acids in the body. Amino acids from all these sources i.e. dietary amino acids, amino acids from tissue break down and synthesized amino acids constitute "amino acid pool" in the body (Fig. 15.1).

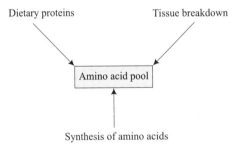

Fig. 15.1   Amino acid pool

Metabolism of proteins involves both synthesis of amino acids (anabolism) as well as breakdown of amino acids (catabolism).

15.1   BIOSYNTHESIS OF AMINO ACIDS

Organisms differ in their abilities to synthesize amino acids. Some bacteria such as *E. coli* are capable of synthesizing all 20 amino acids. Human beings and other animals can synthesize only some of the amino acids. Other amino acids have to be received through diet. Based on this fact, amino acids have been categorized into two types-essential and non-essential amino acids. The amino acids which can be synthesized by the organisms are the non-essential amino acids. Essential amino acids are those amino acids which cannot be synthesized by the organisms and they must be obtained through diet. The pathways for the synthesis of non-essential and essential amino acids are quite different. Non-essential amino acids can be synthesized by quite simple reactions while synthesis of essential amino acids is quite complex.

15.1.1 Biosynthesis of Non-essential Amino Acids

In all the amino acids the source of nitrogen is ammonia. The carbon backbones of amino acids come from glycolytic pathways, pentose phosphate pathway or the citric acid cycle. The α-amino group of most of the amino acids come from α-amino group of glutamate by transamination. All non-essential amino acids except tyrosine are synthesized by any of the following common metabolic intermediates:

Pyruvate, oxaloacetate, α ketoglutarate or 3-phosphoglycerate.

The synthetic pathways of non-essential amino acids are as follows:

Glutamate: Glutamate is synthesized from NH_3 and α-ketoglutarate by the action of enzyme *glutamate dehydrogenase*. The reaction is reversible.

$$\begin{array}{l} COOH \\ | \\ CH_2 \\ | \\ CH_2 \quad + NH_3 + NADPH + H^+ \\ | \\ C{=}O \\ | \\ COOH \end{array} \rightleftharpoons \begin{array}{l} COOH \\ | \\ CH_2 \\ | \\ CH_2 \quad + NADP^+ \\ | \\ CHNH_2 \\ | \\ COOH \end{array}$$

α-Ketoglutarate   Glutamate

Glutamine: Glutamine is formed by further addition of an amino group into glutamate. The reaction is catalyzed by enzyme *glutamine synthetase*. ATP participates directly in the reaction by phosphorylating the side chain of glutamate to form an intermediate compound which reacts with ammonia to form glutamine.

Glutamate γ-glutamylphosphate (intermediate)   Glutamine

Alanine: In several micro-organisms, alanine is synthesized by reductive amination of pyruvate.

$$
\begin{array}{c}
CH_3 \\
| \\
C=O \\
| \\
COOH \\
\text{Pyruvate}
\end{array}
+ NH_3 + NADPH + H^+ \rightleftharpoons
\begin{array}{c}
CH_3 \\
| \\
CHNH_2 \\
| \\
COOH \\
\text{Alanine}
\end{array}
+ NADP^+
$$

In other organisms, alanine is synthesized by transamination reaction. Glutamate is the most important donor of the amino group.

$$
\begin{array}{c}
CH_3 \\
| \\
C=O \\
| \\
COOH \\
\text{Pyruvate}
\end{array}
+
\begin{array}{c}
COOH \\
| \\
CH_2 \\
| \\
CH_2 \\
| \\
CHNH_2 \\
| \\
COOH \\
\text{Glutamate}
\end{array}
\rightleftharpoons
\begin{array}{c}
CH_3 \\
| \\
CHNH_2 \\
| \\
COOH \\
\text{Alanine}
\end{array}
+
\begin{array}{c}
COOH \\
| \\
CH_2 \\
| \\
CH_2 \\
| \\
C=O \\
| \\
COOH \\
\text{α-Ketoglutrate}
\end{array}
$$

The reaction is catalyzed by enzyme *aminotransferase*.

Aspartate: Aspartate is synthesized by transamination of oxaloacetate with glutamate.

$$
\begin{array}{c}
COOH \\
| \\
CH_2 \\
| \\
C=O \\
| \\
COOH \\
\text{Oxaloacetate}
\end{array}
+
\begin{array}{c}
COOH \\
| \\
CH_2 \\
| \\
CH_2 \\
| \\
CHNH_2 \\
| \\
COOH \\
\text{Glutamate}
\end{array}
\rightleftharpoons
\begin{array}{c}
COOH \\
| \\
CH_2 \\
| \\
CHNH_2 \\
| \\
COOH \\
\text{Aspartate}
\end{array}
+
\begin{array}{c}
COOH \\
| \\
CH_2 \\
| \\
CH_2 \\
| \\
C=O \\
| \\
COOH \\
\text{α-Ketoglutarate}
\end{array}
$$

Asparagine: Asparagine is synthesized by asparatate amidation by *asparagine synthetase* in the presence of ATP. ATP is hydrolyzed to AMP and PPi.

$$
\begin{array}{c}
COOH \\
| \\
CH_2 \\
| \\
CHNH_2 \\
| \\
COOH \\
\text{Aspartate}
\end{array}
+
\begin{array}{c}
O=C-NH_2 \\
| \\
CH_2 \\
| \\
CH_2 \\
| \\
CHNH_2 \\
| \\
COOH \\
\text{Glutamine}
\end{array}
\xrightarrow[\text{ATP} \quad \text{AMP+PPi}]{}
\begin{array}{c}
O=C-NH_2 \\
| \\
CH_2 \\
| \\
CHNH_2 \\
| \\
COOH \\
\text{Asparagine}
\end{array}
+
\begin{array}{c}
COOH \\
| \\
CH_2 \\
| \\
CH_2 \\
| \\
CHNH_2 \\
| \\
COOH \\
\text{Glutamate}
\end{array}
$$

Serine: Serine is synthesized from 3-phosphoglycerate. 3-phosphoglycerate is first converted to 3-phosphohydroxypyruvate which undergoes transamination to yield phosphoserine. Phosphoserine is hydrolyzed to serine.

Glycine: Glycine is synthesized by *serine hydroxymethyl transferase* which converts serine to glycine with the yield of N^5, N^{10} methylene-tetrahydrofolate (N^5, N^{10} methylene THF).

Glycine can also be synthesized by condensation of N^5, N^{10}-methylene THF with CO_2 and NH_3 by *glycine synthase*.

Cysteine: Cysteine is formed from serine and homocysteine. Homocysteine is a product of breakdown of methionine. Homocysteine combines with serine to yield cystathionine which forms cysteine and α-ketobutyrate.

$$S - CH_2 - CH_2 - \overset{\overset{\displaystyle H}{|}}{\underset{\underset{\displaystyle NH_2}{|}}{C}} - COOH$$

$$\underset{\underset{\displaystyle NH_2}{|}}{\overset{\overset{\displaystyle H}{|}}{CH_2 - C - COOH}}$$
Cystathionine

$$\longrightarrow \quad H_3C - CH - \overset{\overset{\displaystyle }{}}{\underset{\underset{\displaystyle O}{\|}}{C}} - COOH \quad + \quad HS - CH_2 - \overset{\overset{\displaystyle H}{|}}{\underset{\underset{\displaystyle CH_2}{|}}{C}} - COOH$$

α-Ketobutyrate

Cysteine

Proline: Synthesis of proline occurs from glutamate. In the conversion of glutamate to proline, first glutamate is reduced to form *glutamate-5-semialdehyde* which cyclizes spontaneously to form internal Schiff base *Δ′-pyrroline-5-carboxylate*. This is finally reduced to *proline*.

$$HOOC - CH_2 - CH_2 - \overset{\overset{\displaystyle H}{|}}{\underset{\underset{\displaystyle NH_2}{|}}{C}} - COOH$$
Glutamate

NAD(P)H   NAD(P)$^+$+Pi

ATP   ADP

$$\overset{\overset{\displaystyle O}{\|}}{\underset{\underset{\displaystyle H}{}}{C}} - CH_2 - CH_2 - \overset{\overset{\displaystyle H}{|}}{\underset{\underset{\displaystyle NH_2}{|}}{C}} - COOH$$
Glutamate-5-semialdehyde

Proline

NAD(P)$^+$ NAD(P)H

Δ′–pyrroline-5-carboxylate

Tyrosine: Synthesis of tyrosine is quite a complex process. The precursors are phosphoenol pyruvate and erythrose-4-phospahte which condense to form *2-keto-3-*

$$\underset{\underset{\displaystyle CH_2}{\|}}{\overset{\overset{\displaystyle O - PO_3^{2-}}{|}}{C - COOH}} \quad + \quad \underset{\underset{\displaystyle CH_2OP_3^{2-}}{|}}{\overset{\overset{\displaystyle O \quad H}{\diagdown\!\!/}}{\underset{\displaystyle H - C - OH}{\overset{\displaystyle C}{|}}}}$$

$$\longrightarrow$$

Phosphoenolpyruvate   Erythrose-4-phosphate

$$\overset{\overset{\displaystyle O \quad COOH}{\diagdown\!\!/}}{\underset{\displaystyle }{C}}$$
$$CH_2$$
$$HO - C - H$$
$$H - C - OH$$
$$H - C - OH$$
$$H_2C - O - PO_3^{2-}$$

2-Keto-3-deoxyarabino
heptulosonate-7-phosphate

deoxyarabinoheptulosonate-7-phosphate. This compound cyclizes and ultimately converted to chorismate. Chorismate is converted to *prephenate* and then to tyrosine.

15.1.2 Biosynthesis of Essential Amino Acids

Essential amino acids are synthesized in microorganisms and plants. The pathways for the synthesis of essential amino acids involve more steps as compared to non-essential amino acids.

Synthesis of *phenylalanine* and *tryptophan* occurs from chorismate, which is synthesized from phosphoenolpyruvate and erythrose –4-phosphate. Chorismate is converted to either anthranilate which ultimately forms tryptophan or to prephenate which forms phenylalanine.

In the synthesis of *lysine, methionine* and *threonine*, the common precursor is aspartate. Aspartate is converted to β-aspartate-semialdehyde via aspartyl-β-phosphate as its intermediate. β-aspartate semialdehyde is the common intermediate in the synthesis of methionine, threonine and lysine. β-aspartate semialdehyde undergoes reduction to form homoserine. Homoserine condenses with cysteine to form homocysteine. Homocysteine is converted to methionine by methylation of homocysteine. The reaction is catalyzed by the enzyme *methionine synthase*. The conversion of homoserine to threonine involves two steps. First, – OH group of homoserine is phosphorylated by *homoserine kinase*. Finally, the alcohol group is transferred to secondary position to form threonine.

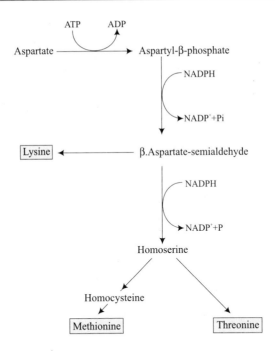

Pyruvate is precursor for the synthesis of *valine* and *leucine*. In the first step, acetaldehyde molecule derived from pyruvate, condenses with another molecule of pyruvate to form α-acetoacetate. α-Acetoacetate forms α-ketoisovalerate which is a precursor for both valine and leucine. α-Ketoisovalerate undergoes transamination to form valine. In the synthesis of leucine, α-ketoisovalerate is first acylated to form β-isopropyl-malate which undergoes oxidative decarboxylation to yield α-ketoisocaproate. α-Ketoisocaproate is transaminated to produce leucine.

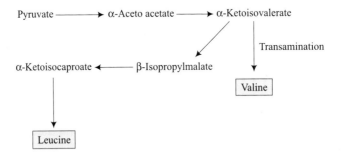

In the synthesis of *isoleucine*, first, α-ketobutyric acid condenses with a molecule of acetaldehyde, derived from pyruvate to produce α-aceto-α-hydroxybutyrate. α-Ketotbutyric acid is derived by deamination of threonine by the enzyme *threonine deaminase*. α-Aceto-α-hydroxybutyrate is then converted to ketoisoleucine which is transaminated to produce isoleucine.

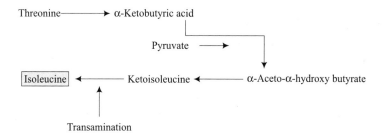

Arginine is synthesized from glutamate. In the process, first glutamate is reduced to form glutamate-5-semialdehyde. Glutamate-5-semialdehyde undergoes transamination to form ornithine. Ornithine is converted to citrulline, utilizing one molecule of carbamoyl phosphate. Citrulline then forms argininosuccinate which is finally hydrolyzed to arginine and fumarate.

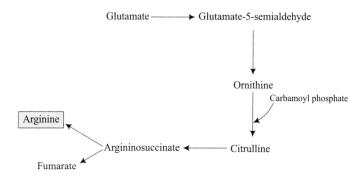

Histidine has a unique biosynthetic pathway. The amino acid is derived from 5-phosphoribosyl-α-pyrophosphate (PRPP) and ATP. Five of the six carbon atoms of histidine are derived from PRPP while the sixth carbon atom is derived from ATP.

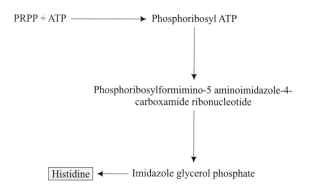

15.2 CATABOLISM OF AMINO ACIDS

In the body amino acids undergo catabolism to form urea and other derivatives which are excreted in the urine. Amino acids can also be used as a source of energy, in case energy supply is not full filled by carbohydrates and fats in the body. Catabolism of amino acids involve the following processes:

  (i) Transamination
 (ii) Deamination

  • Oxidative deamination
  • Non-oxidative deamination

(iii) Decarboxylation
(iv) Urea formation.

Transamination: Transamination is the process in which amino group of one amino acid (donor) is transferred to a α-keto acid (recipient) resulting in the formation of a new amino acid and a new keto acid.

$$
\begin{array}{ccccc}
\text{COOH} & & \text{COOH} & \text{COOH} & \text{COOH} \\
| & & | & | & | \\
\text{CH} - \text{NH}_2 & + & \text{C}=\text{O} \rightleftharpoons & \text{C}=\text{O} & + & \text{CHNH}_2 \\
| & & | & | & | \\
\text{R}_1 & & \text{R}_2 & \text{R}_1 & \text{R}_2 \\
\text{(Donor)} & & \text{(Recipient)} & \text{New keto acid} & \text{New amino acid}
\end{array}
$$

The donor amino acid becomes a keto acid and the recipient keto acid becomes a new amino acid. The process is reversible and is catalyzed by the enzyme known as *transaminase* or *aminotransferase*. This enzyme requires pyridoxal-5′-phosphate as the coenzyme. Pyridoxal-5′-phosphate is a derivative of pyridoxine (vitamin B_6). Transamination takes place principally in liver, kidney, heart and brain. The enzymes are present in almost all mammalian tissues and the process of transamination can be carried out in all tissues to some extent.

Deamination: Deamination is the process in which amino group ($-\text{NH}_2$) is removed from the amino acids to form keto acids. Amino group is removed as ammonia.

$$
\begin{array}{ccc}
\text{R} & & \text{R} \\
| & & | \\
\text{NH}_2 - \text{C} - \text{COOH} & \longrightarrow & \text{O}=\text{C} - \text{COOH} + \text{NH}_3 \\
| & & \\
\text{H} & &
\end{array}
$$

In humans, most of the ammonia is converted into urea in the liver and is eliminated with urine. Deamination usually takes place in liver and kidney. There are two types of deamination:

(i) Oxidative deamination

(ii) Non-oxidative deamination.

(i) *Oxidative deamination*: This process is catalyzed by a group of flavin enzymes, known as *amino acid oxidases*. In the process, the amino acid is first dehydrogenated by the flavo-protein (FP) of the enzyme L-amino oxidase to form α-imino acid. In the next step with addition of H_2O molecule, α-imino nitrogen is released as NH_3 and α-keto acid is formed.

Amino acid oxidases are autooxidizable flavoproteins. Reduced flavoproteins are oxidized to form hydrogen peroxide, which is broken up into H_2O and O_2 by enzyme catalase.

(ii) *Non-oxidative demanination*: There are certain amino acids which can be non-oxidatively deaminated by specific enzymes and form NH_3 e.g. glutamic acid undergoes non-oxidative deamination in the presence of *glutamic dehydrogenase*. NAD acts as a coenzyme.

Sulphur containing amino acids are deaminated by a primary desulph-hydration (removal of H_2S), forming an imino acid which is hydrolyzed to form a keto acid. Pyridoxal phosphate acts as the cofactor.

$$\underset{\substack{\text{L-Cysteine}}}{\underset{\displaystyle \text{COOH}}{\overset{\displaystyle \text{CH}_2-\text{SH}}{\mid}}\!\!\!\overset{\displaystyle \text{CH}-\text{NH}_2}{\mid}} \xrightarrow[\text{H}_2\text{S}]{} \underset{\substack{\text{(Imino acid)}}}{\overset{\displaystyle \text{CH}_2}{\underset{\displaystyle \text{COOH}}{\overset{\displaystyle \parallel}{\text{C}-\text{NH}_2}}}} \longleftrightarrow \underset{\substack{}}{\overset{\displaystyle \text{CH}_2}{\underset{\displaystyle \text{COOH}}{\overset{\displaystyle \parallel}{\text{C}=\text{NH}}}}} \text{(Imino acid)}$$

$$\text{H}_2\text{O}$$

$$\text{NH}_3 \qquad \underset{\boxed{\text{Keto acid}}}{\overset{\displaystyle \text{CH}_2}{\underset{\displaystyle \text{COOH}}{\overset{\displaystyle \parallel}{\text{C}=\text{O}}}}}$$

Decarboxylation: Decarboxylation is the process in which CO_2 is removed from the COOH group of an amino acid and as a result an amine is formed. The reaction is catalyzed by the enzyme *decarboxylase*. Pyridoxal-phosphate acts as the coenzyme.

$$\underset{\boxed{\text{Amino acid}}}{\overset{\displaystyle \text{H}}{\underset{\displaystyle \text{NH}_2}{\overset{\displaystyle \mid}{\text{R}-\text{C}-\text{COOH}}}}} \underset{\text{decarboxylase}}{\longleftrightarrow} \underset{\boxed{\text{Amine}}}{\overset{\displaystyle \text{H}}{\underset{\displaystyle \text{NH}_2}{\overset{\displaystyle \mid}{\text{R}-\text{C}-\text{H}}}}} + CO_2$$

Urea cycle: Excess of ammonia, derived from the catabolism of amino acids is converted to urea which is excreted in the urine. Urea formation is a cyclic process and the cycle is known as *urea cycle*. The urea cycle was outlined by Hans Krebs and Kurt Henseleit in 1932. The cycle is also known as Krebs-Hanseleit cycle. Urea formation occurs in liver and after its formation urea passes into the blood stream and from blood to kidneys and is excreted into the urine.

Urea synthesis takes place in five steps. Each step is catalyzed by a specific enzyme. Out of these five enzymatic reactions, two take place in the mitochondria since their enzymes are present in mitochondria, and other three take place in the cytoplasm. The reactions of urea cycle are as follows (Fig. 15.2):

1. *Synthesis of carbamoyl phosphate*: This is the first reaction of urea formation. The mitochondria enzyme *carbamoyl phosphate synthetase I* catalyzes the condensation of NH_3^+ and HCO_3^- to form carbamoyl phosphate. The phosphate group is derived from ATP. There are two types of carbamoyl phosphate synthetase. Carbamoyl phosphate synthetase I occurs in the mitochondria of liver cells and participates in urea formation. Carbamoyl phosphate synthetase II occurs in cytosol of liver cells and is involved in pyrimidine synthesis.

Carbamoyl phosphate synthetase I is allosterically activated by N-acetylglutamate, which is synthesized from glutamate and acetyl-CoA. When the rate of amino acid breakdown increases, the concentration of glutamate stimulates the synthesis of N-acetylglutamate which activates carbamoyl phosphate synthetase I. As a result, the rate of formation of urea increases.

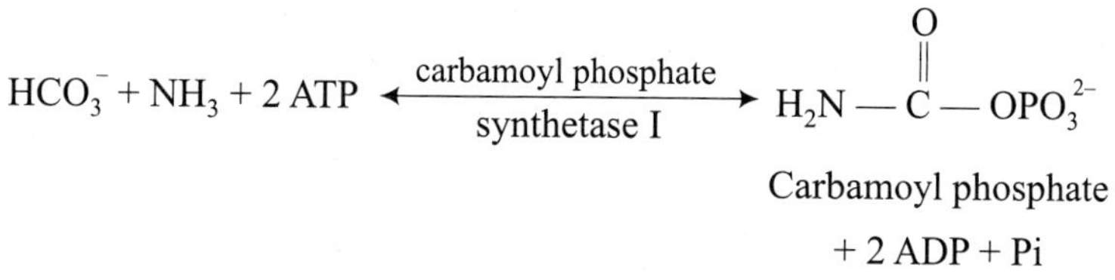

$$HCO_3^- + NH_3 + 2\,ATP \xrightleftharpoons[\text{synthetase I}]{\text{carbamoyl phosphate}} H_2N - \overset{\overset{\displaystyle O}{\|}}{C} - OPO_3^{2-}$$

Carbamoyl phosphate

+ 2 ADP + Pi

2. *Synthesis of Citrulline*: Enzyme *ornithine transcarbamoylase* catalyzes the transfer of carbamoyl group of carbamoyl phosphate to ornithine to yield *citrulline*. Ornithine transcarbamoylase is also a mitochodrial enzyme and thus the reaction occurs inside the mitochondria. As the reaction occurs in mitochodria, ornithine which is produced in the cytosol is transported into the mitochondria. This occurs with the help of specific transport proteins.

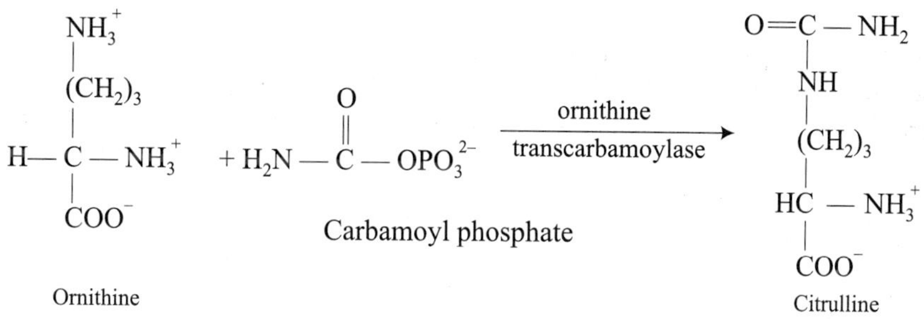

Ornithine + Carbamoyl phosphate → Citrulline (ornithine transcarbamoylase)

3. *Synthesis of Argininosuccinate*: After its formation inside the mitochondria, citrulline is transported to cytosol. In the cytosol citrulline condenses with aspartate to form argininosuccinate. This ATP dependent reaction is catalyzed by *argininosuccinate synthetase*.

Citrulline + Aspartate → Argininosuccinate (argininosuccinate synthetase, ATP → AMP+PPi)

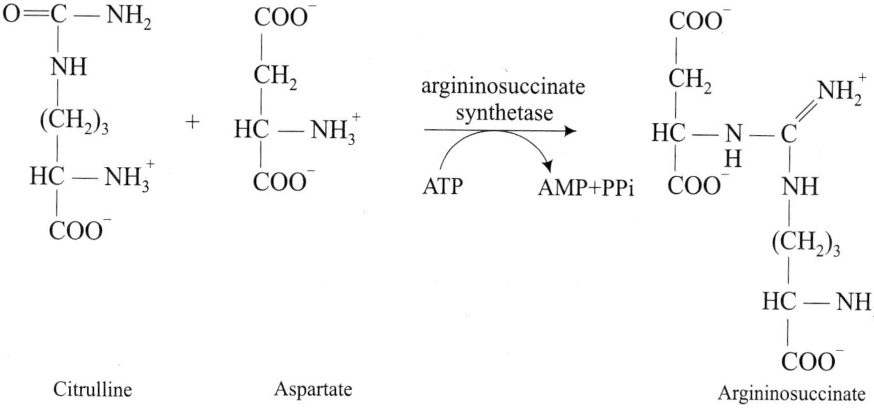

4. *Cleavage of Argininosuccinate*: Argininosuccinate is cleaved into arginine and fumarate by the enzyme *argininosuccinase*. Arginine is the immediate precursor of urea. Fumarate can be reconverted to aspartate for reuse in the argininosuccinate synthetase reaction.

Argininosuccinate   Arginine   Fumarate

5. *Hydrolysis of Arginine*: This is the last reaction of urea cycle. Arginine is hydrolyzed to yield urea and ornithine. The reaction is catalyzed by the enzyme *arginase*. Ornithine is returned to mitochondria for another round of cycle.

Arginine   Urea   Ornithine

15.3   FATE OF AMINO ACIDS

Compounds obtained from the catabolism of amino acids can be further metabolized to form CO_2 and H_2O or can be used in gluconeogenesis. Complete oxidation of amino acids accounts for 10-15% of the metabolic energy generated by the animals. Amino acids can be degraded either to form compounds which are the precursors of glucose or the compounds which are the precursors of fatty acids or ketone bodies. On the basis of their catabolic pathways, the amino acids are divided into three groups:

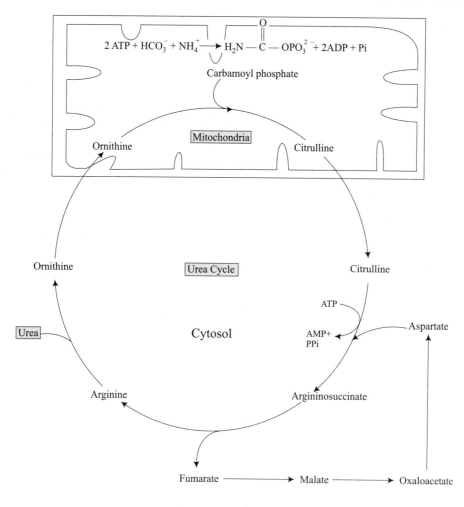

Fig. 15.2 Urea cyle

(i) *Glucogenic amino acids*: These amino acids are degraded to pyruvate, α-ketoglutarate, succinyl-CoA, fumarate or oxaloacetate and are therefore precursors of glucose e.g. alanine, cysteine, glycine, threonine, serine, arginine, glutamate, glutamine, histidine, proline isoleucine, methionine, valine, asparagine, aspartate, phenylalanine, tyrosine, tryptophan are glucogenic amino acids.

(ii) *Ketogenic amino acids* are broken down to acetyl-CoA and acetoacetate and thus can be converted to fatty acid or ketone bodies e.g. leucine, lysine, phenylalanine, tyrosine, isoleucine, tryptophan, threonine.

(iii) *Glucogenic and ketogenic amino acids*: Some amino acids which are the precursors of both carbohydrates and ketone bodies are placed in this category. Phenylalanine, tyrosine, isoleucine, threonine and tryptophan are both ketogenic and glucogenic.

15.4 SUMMARY

- Protein metabolism refers to the synthesis of amino acids (anabolism) and break down of amino acids (catabolism).
- Essential and Non-essential amino acids differ in their biosynthetic pathways.
- Non-essential amino acids are synthesized in simple pathways from pyruvate, oxaloacetate, α-ketoglutarate or 3-phosphoglycerate.
- The pathways for the synthesis of essential amino acids are much more complicated and involve more steps as compared to pathways for the synthesis of non-essential amino acids.
- Degradation of amino acids involves transamination, deamination or decarboxylation.
- Urea synthesis takes place in five steps. Each step is catalyzed by a specific enzyme.

EXERCISE

1. Describe the role of glutamate in the biosynthesis of amino acids.
2. Discuss the biosynthesis of amino acids through transamination.
3. Explain urea cycle.
4. Describe the pathway of biosynthesis of tyrosine.
5. Write notes on:
    (i) Transamination
    (ii) Oxidative deamination

Metabolism of Nucleic Acids

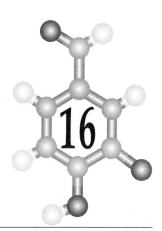

Nucleic acids are the polymers of nucleotides. A nucleotide is composed of a sugar (ribose or deoxyribose), a nitrogenous base and a phosphate group.

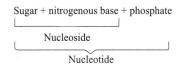

Nitrogenous bases present in DNA and RNA are of two types-purines and pyrimidines. Adenine and guanine are the purines which form adenylate and guanylate nucleotides. Adenylate and guanylate both, are present in RNA. In DNA are present, deoxyadenylate (dAMP) and deoxy guanylate (dGMP). Cytosine, uracil and thymine are the pyrimidines. Cytidylate (CMP) and uridylate (UMP) are present in RNA. In DNA deoxycytidylate (dCMP) and deoxythymidylate (dTMP) nucleotides are present. Thus, there are two types of nucleotides-purine nucleotides and pyrimidine nucleotides.

The metabolism of nucleic acids refers to metabolism of nucleotides.

16.1   BIOSYNTHESIS OF NUCLEOTIDES

The biosynthesis of nucleotides involves two pathways:
1. De novo pathway,
2. Salvage pathway.

   De novo pathway: De novo synthesis begins with the precursors–amino acids, ribose-5-phosphate, CO_2 and NH_3. De novo pathways for purine and pyrimidine nucleotides biosynthesis appear to be nearly identical in all living organisms. Several important precursors are shared by de novo pathways for synthesis of purine and pyrimidine nucleotides.

Salvage pathway: In salvage pathways, free bases and nucleosides, which are released from nucleic acid breakdown, are recycled to form nucleotides.

16.2 DE NOVO BIOSYNTHESIS OF PURINE NUCLEOTIDES

The two purine nucleotides of nucleic acids are: adenosine 5' monophosphate (AMP) and guanosine 5' monophosphate (GMP). The pathway of these purine nucleotide biosynthesis was investigated by Buchanan and G. Robert Greenberg. The pathway of synthesis is essentially same in all living organisms. Purines are first synthesized as nucleotide *inosine monophosphate (IMP)* which is then converted into adenine and guanine nucleotides. Purines are formed as ribonucleotides rather than as free bases. The biosynthesis of purine nucleotides occurs in the following steps (Fig. 16.1).

1. *Formation of PRPP*: The synthesis begins with the formation of 5-*phosphoribosyl-α pyrophosphate (PRPP)* from α-D-ribose-5 phosphate, a product of the pentose pathway. The reaction is catalyzed by the enzyme *ribose phosphate pyrophosphokinase* in the presence of ATP.

2. *Formation of 5-Phosphoribosylamine*: Enzyme *amido phosphoribosyl transferase* catalyzes the displacement of PRPP's pyrophosphate group by amide group of glutamine. This results in the formation of *β-5 phosphoribosylamine*.

3. *Formation of GAR*: Glycine condenses with β-5-phosphoribosylamine to form *glycinamide ribotide* (GAR). The reaction is catalyzed by the enzyme *GAR synthetase* in presence of ATP as the energy source.

4. *Formation of FGAR*: The amino group of GAR is formylated by N^{10}-*formyl H_4-folate* to form *formylglycinamide ribotide* (FGAR). The reaction is catalyzed by the enzyme *GAR transformylase*.

5. *Formation of FGAM*: The amide group of another glutamine is transferred to FGAR to form *formylglycinamidine ribotide* (FGAM). The reaction is catalyzed by enzyme *FGAM synthetase* and ATP provides the energy.

6. *Formation of AIR*: In this step the purine imidazole ring is closed with removal of a molecule of H_2O. This yields *5-aminoimidazole ribotide (AIR)*. The reaction is catalyzed by *AIR synthetase* in the presence of ATP.

7. *Formation of CAIR*: In this reaction there is carboxylation of 5-aminoimidazole ribotide to form *carboxyaminoimidazole ribotide* (CAIR). This carboxylation is unusual in that it does not require biotin but instead uses the bicarbonate, generally present in aqueous solutions. The reaction is catalyzed by *AIR carboxylase*.

8. *Formation of SACAIR*: This is an amide-forming condensation reaction in which aspartate condenses with aminoimidazole carboxylate to yield *5-aminoimidazole-4-(N-succinylocarboxyaamide) ribotide* (SACAIR). The reaction is driven by hydrolysis of ATP by the enzyme *SACAIR synthetase*.

9. *Formation of AICAR*: SACAIR is cleaved by the enzyme *adenylosuccinate lyase* with the release of fumarate to yield *5-aminoimidazole-4-carboxamide ribotide (AICAR)*.

10. *Formation of FAICAR*: In this reaction, the final purine ring atom is acquired through formylation by N^{10}-formyl-THF yielding *5-formaminoimidazole-4-carboxamide ribotide (FAICAR)*. The reaction occurs in the presence of enzyme *AICAR transformylase*.

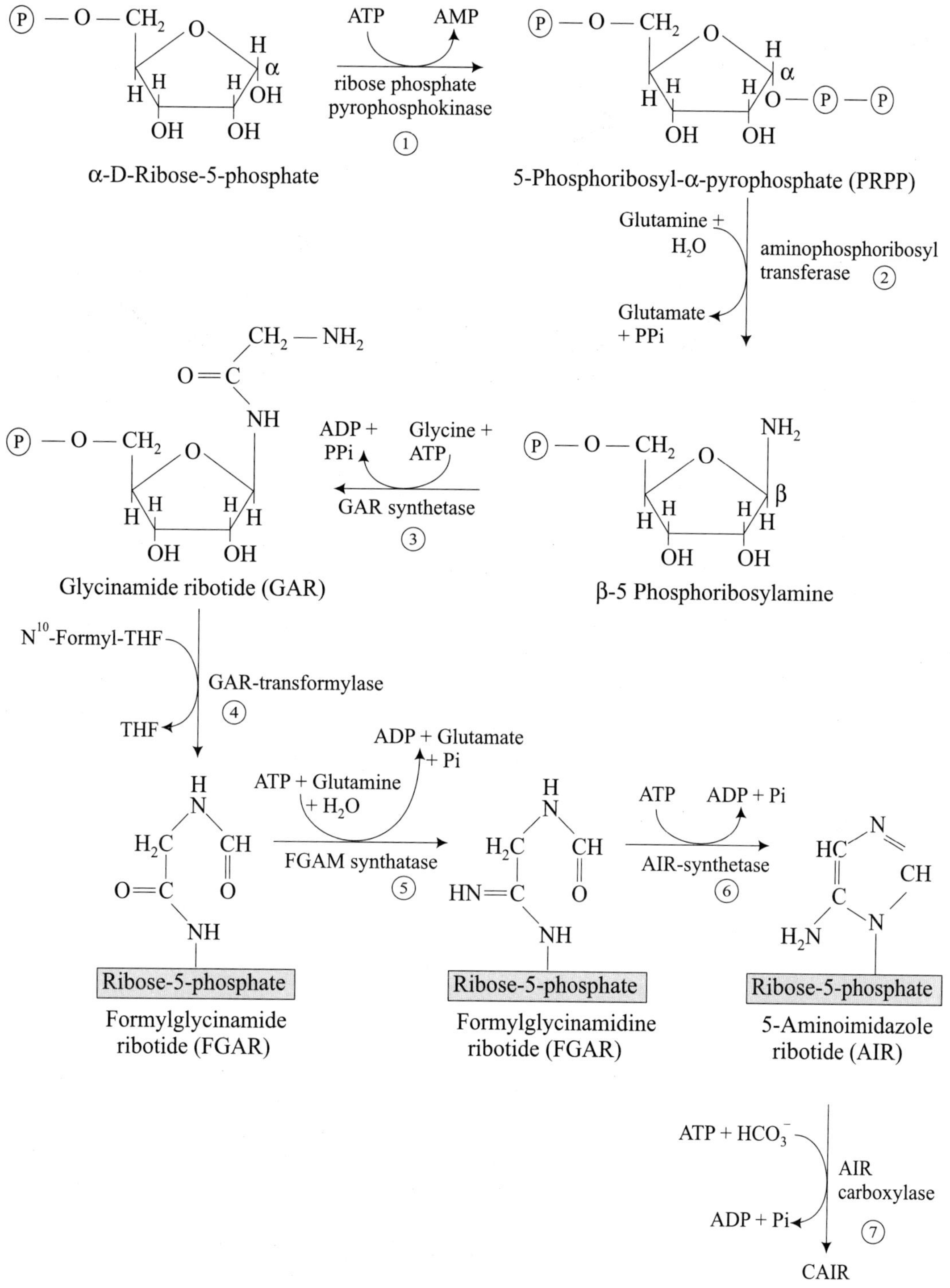

α-D-Ribose-5-phosphate

ATP → AMP

ribose phosphate pyrophosphokinase ①

5-Phosphoribosyl-α-pyrophosphate (PRPP)

Glutamine + H_2O

aminophosphoribosyl transferase ②

Glutamate + PPi

Glycinamide ribotide (GAR)

ADP + PPi ← Glycine + ATP

GAR synthetase ③

β-5 Phosphoribosylamine

N^{10}-Formyl-THF

GAR-transformylase ④

THF

Formylglycinamide ribotide (FGAR)

Ribose-5-phosphate

ATP + Glutamine + H_2O → ADP + Glutamate + Pi

FGAM synthatase ⑤

Formylglycinamidine ribotide (FGAR)

Ribose-5-phosphate

ATP → ADP + Pi

AIR-synthetase ⑥

5-Aminoimidazole ribotide (AIR)

Ribose-5-phosphate

ATP + HCO_3^- → ADP + Pi

AIR carboxylase ⑦

CAIR

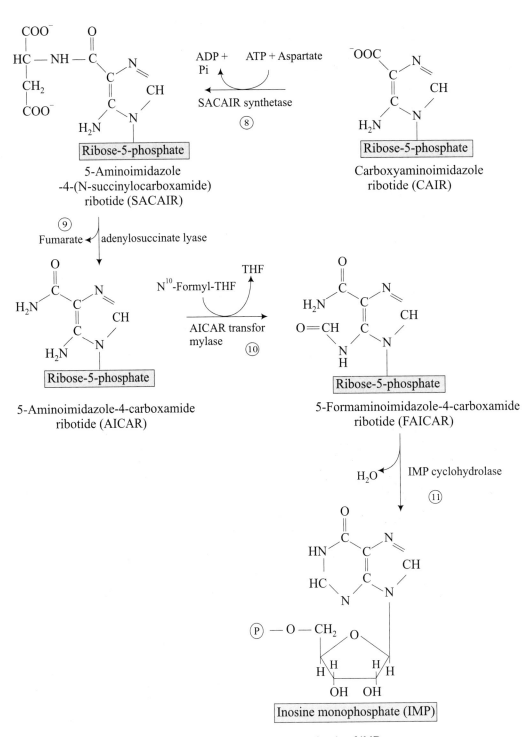

Fig. 16.1 De novo synthesis of IMP

11. *Formation of IMP*: 5-formaminoimidazole-4-carboxamide ribotide undergoes a dehydrative ring closure, by elimination of one molecule of H_2O to form inosine monophosphate (IMP). The reaction is catalyzed by the enzyme *IMP cyclohydrolase*.

Synthesis of AMP and GMP: Inosine monophospate (IMP) does not accumulate in the cell and is rapidly converted into AMP and GMP (Fig. 16.2).

Formation of AMP: The synthesis of AMP takes place in two steps:

In the first step, enzyme *adenylosuccinate synthetase* catalyzes the condensation of Asparate with IMP to form *adenylosuccinate*. In this reaction energy is provided by GTP.

Adenylosuccinate is then cleaved to form AMP and fumarate by enzyme *adenylosuccinate lyase*.

Formation of GMP: Formation of GMP from IMP also occurs into two steps:

In the first step, IMP is dehydrogenated to form *xanthosine monophosphate* (XMP) by the enzyme *IMP dehydrogenase*.

XMP is then converted to GMP by the transfer of the glutamine amide nitrogen in a reaction driven by the hydrolysis of ATP to AMP + PPi.

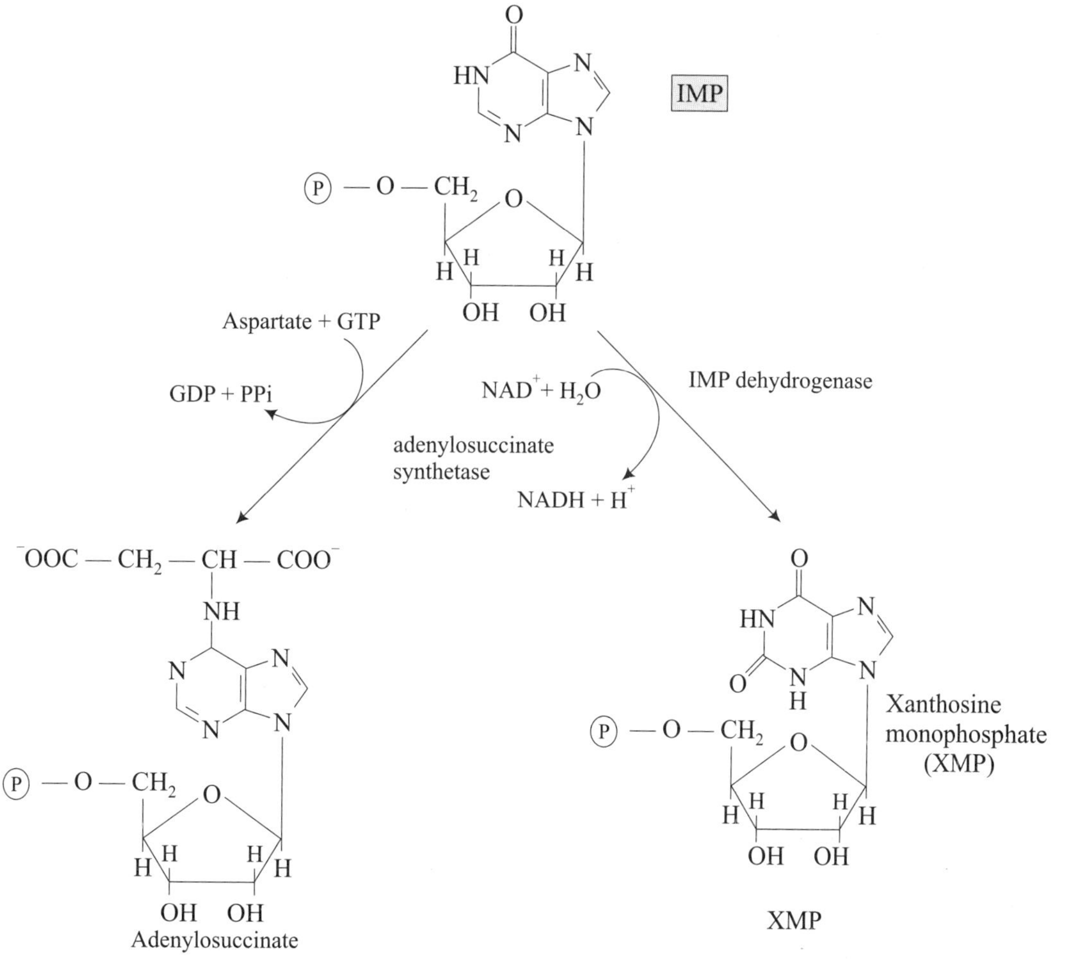

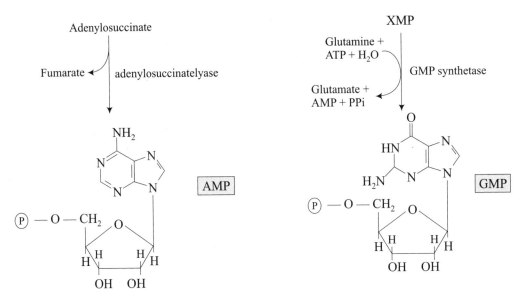

Fig. 16.2 Synthesis of AMP and GMP from IMP

16.3 CATABOLISM OF PURINES

The catabolism of purines yield uric acid as the end-product. Adenine and guanine nucleotides have their separate enzymes which lead to the formation of common product xanthine. Finally, xanthine is converted to uric acid by *xanthine oxidase*. Uric acid is the excreted end product of purine catabolism in primates, birds and many other animals (Fig. 16.3).

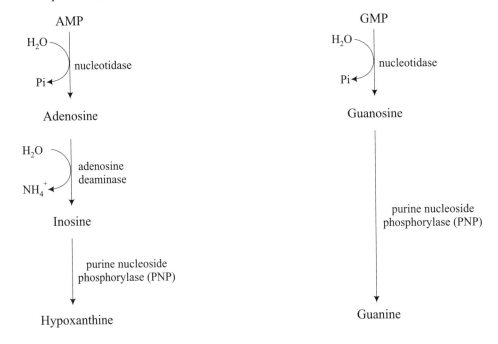

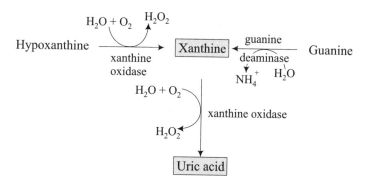

Fig. 16.3 Purine catabolism

16.4 BIOSYNTHESIS OF PYRIMIDINE NUCLEOTIDES

De nove biosynthesis of pyrimidine nucleotides is different from purine nucleotide synthesis. Here, the six-membered pyrimidine ring is made first-and then attached to ribose-5 phosphate. Uridine 5′ monophosphate (UMP, Uridylate) is synthesized in a six-reaction pathway. UMP is the precursor of CMP i.e. cytidine 5′ monophosphate (Cytidylate).

Synthesis of UMP involves the following steps (Fig. 16.4).

- *Synthesis of carbamoyl phosphate*: The synthesis of pyrimidine ring begins with the synthesis of carbamoyl phosphate from glutamine, CO_2 and ATP. The reaction is catalyzed by the enzyme *carbamoyl phosphate synthetase II* present in the cytosol.

- *Synthesis of carbamoyl aspartate*: Carbamoyl phosphate reacts with aspartate to yield *carbamoyl aspartate*. The reaction is catalyzed by *aspartate transcarbamoylase*.

- *Formation of dihydroorotate*: By removal of water from carbamoyl aspartate, the pyrimidine ring is closed to form *dihydroorotate*. The reaction is catalyzed by enzyme *dihydroorotase*.

- *Formation of orotate*: Dihydroorotate is oxidized to orotate by the enzyme *dihydroorotate dehydrogenase*. NAD^+ acts as the electron acceptor.

- *Formation of orotidylate*: Orotate reacts with PRPP to yield *orotidine 5′ monophospate (orotidylate)*. The reaction is catalyzed by enzyme *orotate phosphoribosyl transferase*.

- *Formation of UMP*: Orotidine 5′ monophospate is decarboxylated to yield *uridine monophosphate (UMP)* by enzyme *OMP decarboxylase*.

UMP is phosphorylated by ATP to form UDP (uridine diphosphate) which can be further phosphorylated to form UTP (uridine triphosphate). The reactions are catalyzed by *nucleoside monophosphokinase*.

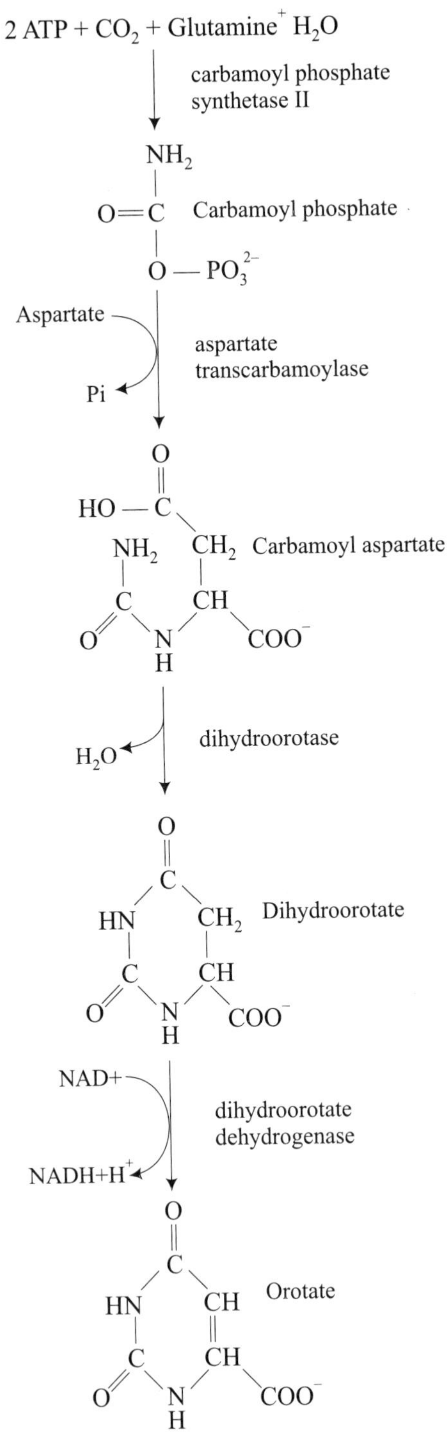

$$2\ ATP + CO_2 + Glutamine^+\ H_2O$$

carbamoyl phosphate synthetase II

$$
\begin{array}{c}
NH_2 \\
| \\
O{=}C \qquad \text{Carbamoyl phosphate} \\
| \\
O - PO_3^{2-}
\end{array}
$$

Aspartate

aspartate transcarbamoylase

Pi

Carbamoyl aspartate

$$
\begin{array}{c}
O \\
\parallel \\
HO - C \\
NH_2 \qquad CH_2 \\
C \qquad CH \\
O \qquad N \qquad COO^- \\
H
\end{array}
$$

H_2O \qquad dihydroorotase

Dihydroorotate

$$
\begin{array}{c}
O \\
\parallel \\
C \\
HN \qquad CH_2 \\
C \qquad CH \\
O \qquad N \qquad COO^- \\
H
\end{array}
$$

NAD+

dihydroorotate dehydrogenase

$NADH{+}H^+$

Orotate

$$
\begin{array}{c}
O \\
\parallel \\
C \\
HN \qquad CH \\
\qquad \parallel \\
C \qquad CH \\
O \qquad N \qquad COO^- \\
H
\end{array}
$$

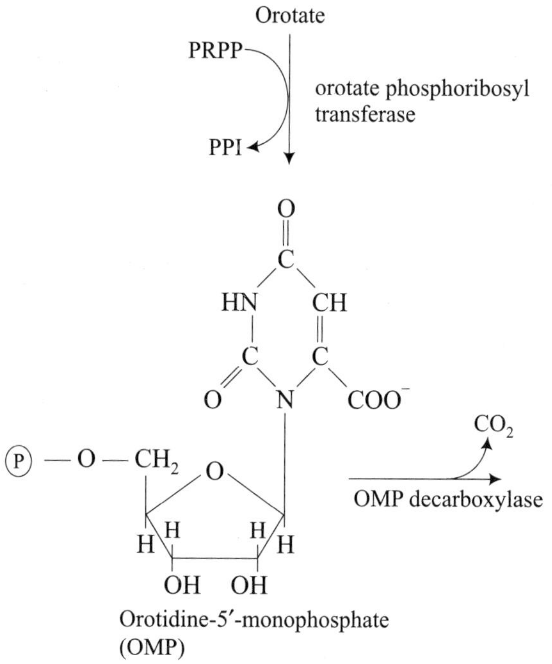

Fig. 16.4 De novo synthesis of UMP

Synthesis of CTP: CTP is synthesized from UTP by its amination. The reaction is catalyzed by the enzyme *CTP synthetase*. The amino group is provided by glutamine in the presence of ATP.

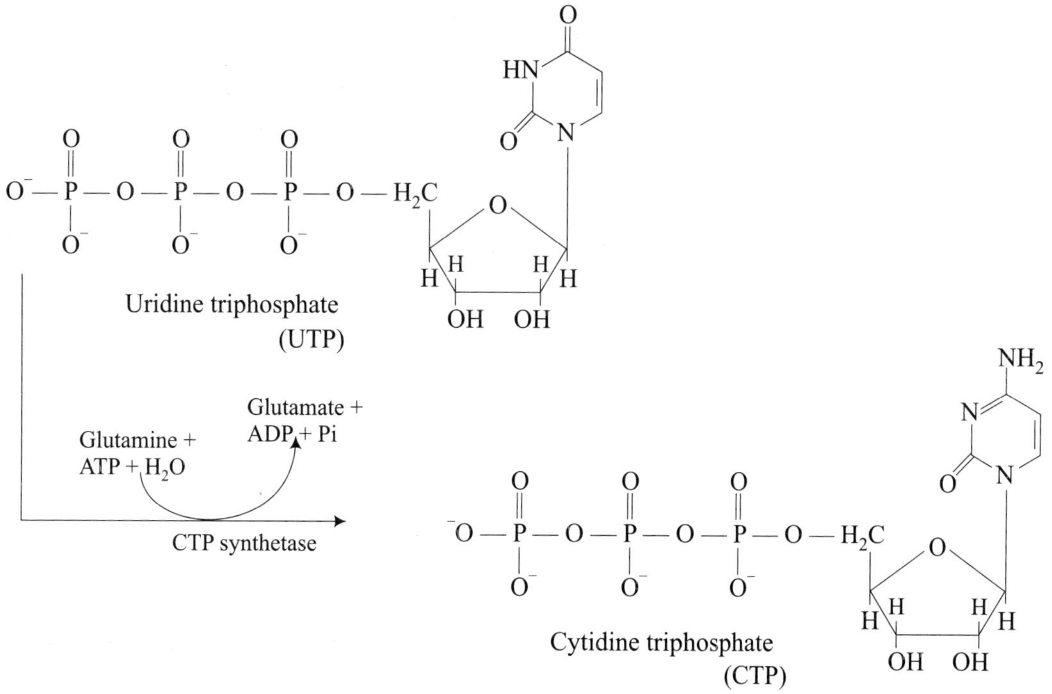

Formation of deoxyribonucleotides: Deoxyribonucleotides are derived from corresponding ribonucleotides by direct reduction at the 2′ carbon atom of ribose to form 2′ deoxy derivative. The reaction is catalyzed by *ribonucleotide reductase*. The reaction requires thioredoxin and NADH.

16.5 CATABOLISM OF PYRIMIDINES

Pyrimidine nucleotides are degraded to their component bases. These bases are then broken down in the liver through reduction. The end products are β-alanine (uracil and cytosine) and β-amino isobutyrate (thymine). These amino acids are converted to malonyl-CoA and methylmalonyl-CoA through transamination and activation reactions. Malonyl-CoA is a precursor of fatty acid synthesis and methylmalonyl-CoA is converted to succinyl-CoA cytidine (Fig. 16.5).

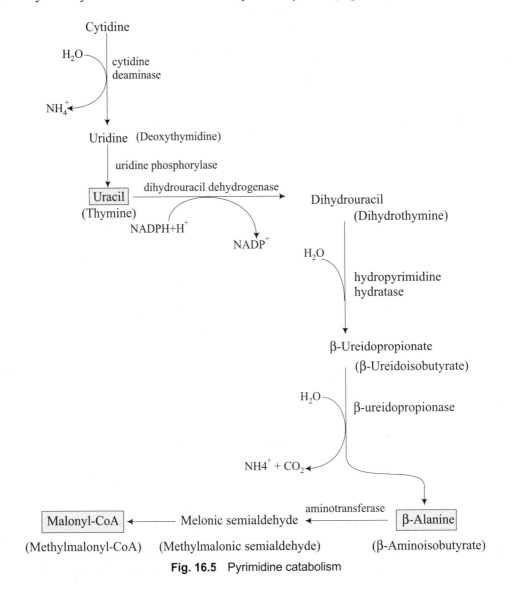

Fig. 16.5 Pyrimidine catabolism

16.6   SALVAGE PATHWAYS FOR PURINES AND PYRIMIDINE BASES

Free purines and pyrimidines are released in cells as a result of metabolic degradation of nucleic acids. These are reused to make nucleotides. In mammals purines are salvaged by two different enzymes:

Adenine phosphoribosyl transferase (*APRT*) reacts with PRPP to yield adenine nucleotide.

$$\text{Adenine} + \text{PRPP} \xrightarrow{\text{APRT}} \text{AMP} + \text{PPi}$$

Similarly, *Hypoxanthine-guanine phosphoribosyl transferase* (HGPRT) catalyzes the formation of GMP and IMP from guanine and hypoxanthine.

$$\text{Guanine} + \text{PRPP} \xrightarrow{\text{HGPRT}} \text{GMP} + \text{PPi}$$

$$\text{Hypoxanthine} + \text{PRPP} \xrightarrow{\text{HGPRT}} \text{IMP} + \text{PPi}$$

In the same way, *pyrimidine phosphoribosyl transferase* catalyzes the formation of pyrimidine nucleotide using PRPP as the donor of ribosyl moiety.

$$\text{Pyrimidine base} + \text{PRPP} \xrightarrow{\text{PPRT}} \text{Pyrimidine nucleotide} + \text{PPi}$$

16.7   SUMMARY

- Nucleic acid is composed of monomeric units called nucleotides.
- A nucleotide is composed of a sugar, a base and a phosphoric acid molecule.
- Two types of pathways lead to the synthesis of nucleotides–de novo pathways and salvage pathways.
- Synthesis of purine nucleotides occurs in 11 steps which results in the formation of IMP.
- IMP forms AMP and GMP.
- The catabolism of purines yields uric acid as the end product.
- Pyrimidine nucleotide UMP is synthesized in 6 steps. UMP is converted to UTP and CTP by phosphorylation and amination.
- Degradation of pyrimidines results in the formation of β-alanine and β-aminoisobutyrate.

EXERCISE

1. Explain the de novo biosynthesis of purine nucleotides.
2. Compare the pathways of purine and pyrimidine nucleotides biosynthesis.
3. Explain:
   (i) Salvage pathway for synthesis of purine nucleotides
   (ii) Catabolism of pyrimidines.

Inborn Errors of Metabolism

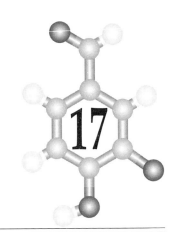

Inborn errors of metabolism is a group of genetic disorders of metabolism. The majority of these disorders are caused by defect in a single gene which codes for a specific enzyme. The defect in the gene causes a defect in structure/function of the enzyme coded by it. This results in abnormality in the metabolic pathway in which this enzyme participates. The defect in the metabolic pathway results either in accumulation of substrate which may be toxic or deficiency of product (essential compounds). In both cases the effect is an abnormal condition. Inborn errors of metabolism are also referred to as *congenital metabolic diseases* or *inherited metabolic diseases*.

Nearly every metabolic disease has several forms that vary in age of onset, clinical severity and mode of inheritance. Such diseases can affect any organ system and usually affect multiple organ systems. Manifestations vary from those of acute life-threatening disease to subacute progressive degenerative disorder.

Depending on the metabolic pathway, in which the defect occurs, main categories of inborn errors of metabolism are as follows:

- Disorders of carbohydrate metabolism e.g. galactosemia, glycogen storage disease.
- Disorders of protein metabolism e.g. phenylketonuria, urea cycle defects.
- Disorders of fatty acid oxidation and mitochondrial metabolism e.g. medium chain acyl-CoA dehydrogenase deficiency.
- Lysosomal storage disorders e.g. Gaucher's disease.
- Disorders of purine or pyrimidine metabolism e.g. Lesch-Nyhan syndrome.
- Disorders of steroid metabolism e.g. congenital adrenal hyperplasia.

Some common inborn errors of metabolism are:

17.1 GLYCOGEN STORAGE DISEASE

Glycogen storage disease is a common disorder of carbohydrate metabolism. The disease results from deficiency of the enzyme *glucose-6-phosphatase*. The enzyme is involved in the formation of glucose from glycogen and from gluconeogenesis. This impairs the ability of the liver to produce free glucose which causes severe hypoglycemia. Reduced glycogen breakdown results in increased glycogen storage in liver and kidneys causing their enlargement.

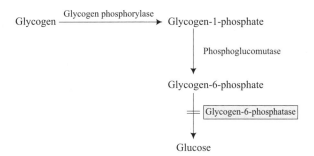

The disease is inherited as an autosomal recessive disease. There are various types of glycogen storage disease. *Type 1* or *Von Gierke's disease* is the most common of the glycogen storage diseases which results from deficiency of enzyme glucose-6-phosphatase. Some common types of glycogen storage disease are given in Table 17.1.

Table 17.1   Glycogen storage disease

Type	Name	Deficient Enzyme	Clinical features
1.	Von Gierke's Disease	Glucose-6-phosphatase	Hypoglycemia, increased levels of uric acid, Hepatomegaly
2.	Pompe's Disease	Acid maltase	Cardiomegaly, Muscle hypotonia, No hypoglycemia.
3.	Forbe's Disease	Debranching enzyme	Hepatomegaly, Moderate hypoglycemia, Acidosis.
4.	Andersen's Disease	Branching enzyme	Hepato-splenomegaly, moderate hypoglycemia, nodular cirrhosis of liver, Hepatic failure.
5.	McArdle's Disease	Muscle phosphorylase	Muscle cramps on exercise, weakness and stiffness of muscle.
6.	Her's Disease	Liver phosphorylase	Hepatomegaly, mild to moderate hypoglycemia, mild acidosis.

17.2   GALACTOSEMIA

Galactosemia is also a disorder of carbohydrate metabolism. The disease is caused by the deficiency of enzyme *galactose-1-phosphate uridyl transferase*. The enzyme takes part in the breakdown of galactose in the body. During its metabolism, galactose is first converted to

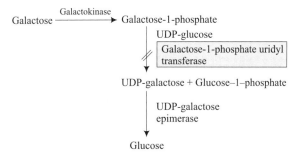

galactose-1-phospahte by galactokinase. Galactose-1-phosphate, in the presence of enzyme galactose-1-phosphate uridyl transferase acquires a uridyl group from uridine disphosphate glucose (UDP-glucose) to form UDP-galactose and glucose-1-phosphate. The galactose moiety of UDP-galactose is then epimerized to glucose. Glucose-1-phosphate is isomerized to Glucose-6-phosphate by phosphoglucomutase.

Due to deficiency of the enzyme galactose-1-phosphate uridyl transferase, galactose is not metabolised. The blood galactose level is markedly elevated and galactose is found in urine. Besides blood, galactose also accumulates in liver, spleen, kidney, heart, lens of eye and cerebral cortex. This can lead to liver damage, kidney failure, mental retardation and cataract in eyes. The disease is inherited as an autosomal recessive disease.

17.3 PHENYLKETONURIA

Phenylketonuria is an inborn error of protein (amino acid) metabolism. The disease is caused due to absence of *phenylalanine hydroxylase* enzyme activity which converts dietary phenylalanine to tyrosine. Due to deficiency of this enzyme phenylalanine cannot be converted to tyrosine and it accumulates in blood. Phenylalanine undergoes transamination to form phenyl pyruvic acid and its products as phenyl lactic acid and phenyl acetic acid. Accumulation of phenylalanine and its products may cause defective formation of serotonin and melatonin. This may also lead to mental retardation.

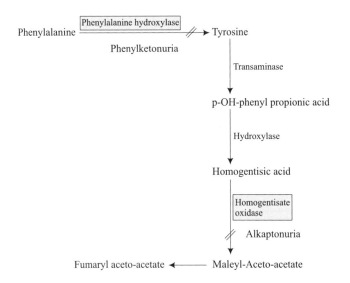

17.4 ALKAPTONURIA

Alkaptonuria is a hereditary defect in the metabolism of amino acid. Its transmission is autosomal recessive form. The disease is caused by the deficiency of the enzyme *homogentisate oxidase*. As a result homogentisic acid accumulates in the tissues and blood and appears in urine. Homogentisic acid is readily oxidized to black pigment (alkapton). Thus, the urine of the affected person turns black when exposed to air. Deposition of homogentisic acid derivatives in connective tissues and cartilages leads to *ochronosis*, characterized by arthritis and cartilage degeneration.

17.5 UREA CYCLE DEFECTS

Urea cycle is an important pathway for the removal of highly toxic ammonia from the body. The cycle involves five different enzymes. Absence of any of these enzymes may lead to the condition of hyperammonemia. (Table 17.2). Inherited disorders due to inherited deficiency of enzymes of urea cycle have been reported. *Hyperammonemia Type I* is due to deficiency of enzyme carbamoyl-P-synthetase I which leads to ammonia toxicity. *Hyperammonemia Type II*, also called as *Ornithinemia* is caused by the deficiency of ornithine transcarbamoylase and characterized by increase in blood ammonia and ornithine. Deficiency of enzyme argininosuccinate synthetase leads to the disease *Citrullinemia. Argininosuccinic aciduria* is caused by the deficiency of argininosuccinase. The condition is characterised by hyperammonemia, ammonia toxicity and mental retardation. *Hyperargininemia*, caused by the deficiency of arginase, manifests as hyperammonemia.

Table 17.2   Urea cycle defects

Defect	Enzyme
1. Hyperammonemia Type I	Carbamoyl-P-synthetase I
2. Hyperammonemia Type II	Ornithine transcarbamoylase
3. Citrullinemia	Argininosuccinate synthetase
4. Argininosuccinic aciduria	Argininosuccinase
5. Hyperargininemia	Arginase

17.6 GAUCHER'S DISEASE

Gaucher disease is a rare inherited disease which belongs to a group of diseases called lysosomal storage disorders. The disease is inherited in an autosomal pattern. Gaucher disease is caused by the deficiency of enzyme *glucocerebrosidase*. The deficiency of the enzyme results in a build up of the glycolipid-glucocerebroside throughout the body-in the bones, liver, spleen, central nervous system and other body organs. Accumulation of glucocerebroside is toxic and results in progressive and permanent damage. There are three types of Gaucher disease. Each type has a characteristic age of onset and constellation of symptoms. Type I is the most common and shows the mildest symptoms. It is non-neurogenic and affects both children and adults. Type II usually begins to show symptoms during infancy and accounts for less than 1% of patients with Gaucher disease. It results in severe and progressive neurological problems. Type III usually occurs during childhood and shows symptoms of both adult and infantile form.

17.7 MEDIUM-CHAIN ACYL-CoA DEHYDROGENASE DEFICIENCY

It is an inborn disorder of lipid metabolism. The disease is due to the deficiency of enzyme *medium chain acyl CoA dehydrogenase (MCAD)*. Individuals with MCAD deficiency cannot metabolize medium-chain fatty acids. Impairment of fatty acid oxidation may lead to hypoglycemia and hyperammonemia. Partially degraded fatty acids may build up in tissues and can damage the liver and brain, causing serious complications.

17.8 CONGENITAL ADRENAL HYPERPLASIA

Congenital adrenal hyperplasia is a condition in which there is impairment of the steroidogensis of cortisol by adrenal gland. Insufficient cortisol production results in rising level of ACTH which causes the hyperplasia of adrenal cortex. Along with cortisol, the synthesis of mineralocorticoids and androgens is also affected. The resulting excessive or deficient production of these hormones leads to a variety of problems. Specific enzyme insufficiencies are associated with over-or underproduction of specific hormone. But in most of the forms of congenital adrenal hyperplasia, there is deficiency of enzyme *21-hydroxlase*. Milder forms of 21-hydroxylase deficiency can cause androgen effects and infertility while in its severe form deficiency of 21-hydroxylase may cause severe dehydration leading to death.

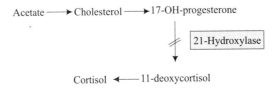

17.9 LESCH-NYHAN SYNDROME

Lesch-Nyhan syndrome is an inborn metabolic error which is inherited as sex-linked recessive disorder. The condition arises due to the absence of enzyme *hypoxanthine-guanine phosphoribosyl transferase*, which is involved in the salvage pathway of purine synthesis. The absence of the enzyme leads to an elevated concentration of PRPP (Phosphoribosyl-2-pyrophosphate), a marked increase in the rate of purine biosynthesis by the de novo pathway and an over production of uric acid. The disorder is characterised by a compulsive self destructive behaviour in affected people. Elevated levels of uric acid in the serum may lead to the formation of kidney stone. Neurological abnormalities such as spasticity and mental retardation are other characteristics of Lesch-Nyhan syndrome.

17.10 SUMMARY

- Inborn errors of metabolism are inherited disorders of metabolism.
- Most of these disorders are due to defect in a single gene.
- Inborn errors of metabolism are inherited in autosomol recessive form.
- These disorders can be treated by nutritional therapy. Appropriate dietary restriction and modifications can be helpful in the treatment of some of these diseases.

EXERCISE

1. What are inborn errors of metabolism? Discuss their characteristic features.
2. Describe different categories of inborn metabolic disorders with examples.
3. Write short notes on:
    (i) Galactosemia
    (ii) Phenylketonuria
    (iii) Gaucher disease

Part III

Gene Concept and Gene Expression

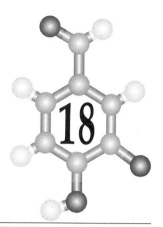

Gene Concept

Mendel studied the inheritance of characters in pea plants. To explain his results he presumed that a character was determined by a pair of 'factors' and factors were the carrier of genetic characters from parent to the offspring. Later, Johannsen (1909) used the term 'gene' for Mendelian 'factors'.

A gene is the hereditary unit which determines structural and functional characteristics in an organism. It exerts its effect through RNA or protein product. In 1910, T.H. Morgan showed that genes are situated on chromosomes and occupy specific locations in the chromosomes. The region of the chromosome in which a particular gene is located is called its *locus*. In the chromosomes genes are arranged in a linear order. Genes can undergo sudden changes in their structure (mutation). The changed gene is called a *mutant gene*.

In 1941, Beadle and Tatum proposed 'one gene one enzyme' hypothesis i.e. a gene codes for a specific protein (enzyme). Mutation in the gene may cause error in its expression and synthesis of a defective enzyme.

A gene may occur in more than one form. These different functional forms of a gene are called *alleles*. Usually a gene has two alleles. But some genes have more than two alleles. These are known as multiple alleles.

18.1 MOLECULAR STRUCTURE OF GENE

At the molecular level, a gene is a segment of DNA which is necessary to form a polypeptide chain or RNA. Most of the genes express themselves by producing mRNA (transcription) which then directs the synthesis of a specific protein (translation). Such genes are called structural genes or protein-coding genes. There are certain other genes which code for tRNA (transfer RNA), rRNA (ribosomal RNA) and SnRNA (small nuclear RNA) molecules. These genes are different from structural genes because their RNA transcripts are the final products. They are directly used and are not translated into proteins. These molecules (RNA/proteins) which are produced by a gene expression are known as gene products.

All the regions of DNA in a gene do not code for protein synthesis. Though these regions do not code for proteins but they are required for the synthesis of proteins. Such regions are called *regulatory regions. Promoters* and *enhancers* are the regulatory regions which do not code for a protein/RNA but control the transcription and RNA processing. Mutations in these regions will affect the normal expression and function of RNA.

The coding region of a gene produces mRNA which may be either monocistronic or polycistronic. A cistron is a genetic unit which encodes a single polypeptide. In prokaryotes mRNA may code for several proteins that function together in a biological process. Such mRNAs are called *polycistronic*. In eukaryotes, mRNAs are *monocistronic* i.e. each mRNA encodes a single protein.

Though genes are the segments of DNA but most of the genes of multicellular organisms are not a continuous sequence of nucleotides but are interrupted by intervening sequences which are not represented in mRNA and are not utilized in protein synthesis. These intervening regions which do not participate in protein synthesis are called *introns* (Fig. 18.1). These are removed from mRNA in a process called *splicing* during RNA processing in the nucleus before the fully processed mRNA reaches the cytoplasm for translation.

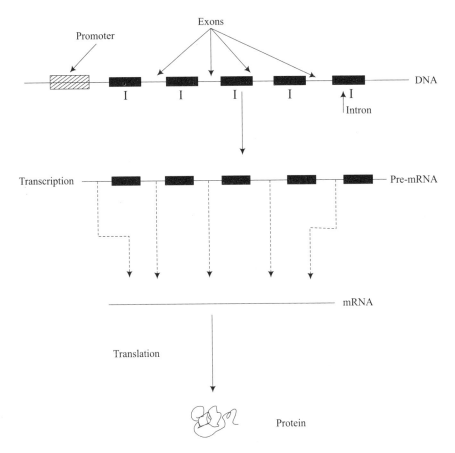

Fig. 18.1 Structure of a typical eukaryotic protein coding gene

The sequences of the gene which actually encode mRNA or amino acid sequence in a protein are called *exons*. An exon is the functional unit of DNA. The intron sequences of genes are often significantly longer than the exons. In case of human, a large portion of precursor mRNA (pre-mRNA) is intronic in nature and is removed by RNA splicing. Prokaryotic cells generally lack introns. The gene containing a single uninterrupted stretch of DNA is called a cistron.

18.2   CHROMOSOMAL ORGANIZATION OF GENE AND NON-CODING DNA

The total sets of genes in an organism is known as *genome*. The number of genes in prokaryotes is less than that in eukaryotes. But within eukaryotes there is no direct relationship between number of genes and the complexity of the organism. In case of human genome, number of genes is about 20,000 – 25,000. There are organisms which have larger number of genes than human beings.

Genes are located on particular regions in a chromosome. These regions are called 'loci'. In an organism since the number of genes is very large in comparison to the number of chromosomes, it is assumed that several genes are located on a single chromosome. In a chromosome genes are located in a linear order. The distribution of genes is not uniform within a chromosome. The density of genes varies greatly in different regions of chromosomal DNA.

In prokaryotes, there is a single circular chromosome which contains most of the genes. Hence, the chromosome has high gene density. Prokaryotes also have extra chromosomal DNA known as plasmid. Plasmids also contain some genes.

In eukaryotes, chromosomes are present in a well defined nucleus. Each chromosome consists of thousands of genes. In eukaryotic DNA there are regions which do not code for protein synthesis. These non-coding regions may have regulatory function in protein synthesis. As discussed earlier, promoters and enhancers are the regulatory regions, present in DNA. Very large stretches of repetitive sequences are present at the ends of chromosomes, known as telomeres. These sequences prevent degradation of coding and regulatory regions during DNA replication.

In most of the organisms, it is only a small part of DNA which actually participates in protein synthesis. In human only about 1.5% of total DNA actually encodes protein or functional mRNA. A large amount of DNA has no obvious function.

18.3   SUMMARY

- A gene is the hereditary unit. It is a unique sequence of nucleotides which encodes a polypeptide chain or RNA.
- A gene occurs in different functional forms which are known as alleles.
- Genes are located on chromosomes in a linear order.
- A gene may include non-transcribed and non-translated sequence that may have regulatory functions.
- Exons are the expressed sequences of gene.
- Introns are the intervening sequences which are not represented in the mature mRNA and do not take part in translation.
- In an organism, only a small part of DNA participates in protein synthesis.

EXERCISE

1. Define a gene. Enlist its salient features.
2. Describe the structure of a gene at the molecular level.
3. Explain the following:
   (i) Regulatory sequences
   (ii) Non-coding DNA

Replication and Transcription of DNA

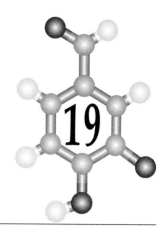

19.1 REPLICATION OF DNA

One of the important characteristics of DNA is its capacity to replicate. Various models have been proposed to explain DNA replication such as:

(i) Conservative model

(ii) Dispersive model

(iii) Semiconservative model

- *Conservative model*: According to conservative model, two strands of parental DNA remain together and serve as a template for the synthesis of new DNA. Thus, out of the two daughter DNA molecules, one is actually the parental DNA while the other molecule consists of totally newly synthesized double strands.

- *Dispersive model*: In the dispersive model, the parental double helix DNA breaks into double stranded DNA segments. These segments act as templates for the synthesis of new double stranded DNA segments which reassemble and form a complete DNA with parental and progeny DNA segments interspersed.

- *Semiconservative model*: According to this model DNA replication takes place by semi-conservative method. Two strands of the parental DNA separate from each other and each strand acts a template for the synthesis of complementary strand of DNA. The sequence of the new strand is determined by the base sequence of template strand. The newly synthesized DNA would have one strand from the parent DNA and the other one–a newly synthesized strand. As the daughter DNA consists of one parental strand and one newly synthesized strand, this model of replication is called semiconservative model. This is a well accepted model of DNA replication. Experimental evidences are there in support of this model (Fig. 19.1).

Meselson-Stahl's experiment in support of semiconservative model

In their experiment, Mathew Meselson and Frank Stahl cultured bacteria *E. coli* in a medium containing N^{15} instead of normal N^{14}. As a result both strands of DNA of all the bacteria contained

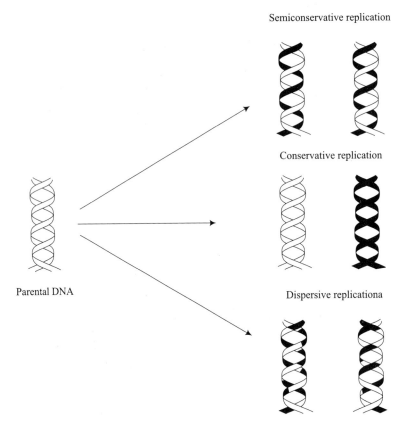

Fig. 19.1 Different models of replication in DNA.

N^{15} as constituent of purines and pyrimidines. N^{15} DNA can be separated from N^{14} DNA by using equilibrium density gradient centrifugation.

These bacteria with N^{15} were then transferred to a medium containing N^{14} and were allowed to replicate in the new N^{14} medium. After one generation in the medium, it was found that all DNA of bacteria had a density that was intermediate between totally N^{15} DNA and totally N^{14} DNA. After two generations half of the DNA was of intermediate density and half was of the density of N^{14} containing DNA. The DNA with intermediate density indicates that the newly synthesized DNA has one heavier strand (parental DNA) and other lighter one (newly synthesized DNA). These results clearly support the semiconservative model of DNA replication (Fig. 19.2).

19.1.1 Mechanism of Replication

Replication begins with the relaxation of supercoiled DNA. Supercoiled DNA is relaxed by *topoismerase*. This is followed by the separation of two strands of DNA. The unwinding of the parental DNA begins at unique segments in a DNA molecule called *replication origins* or simply

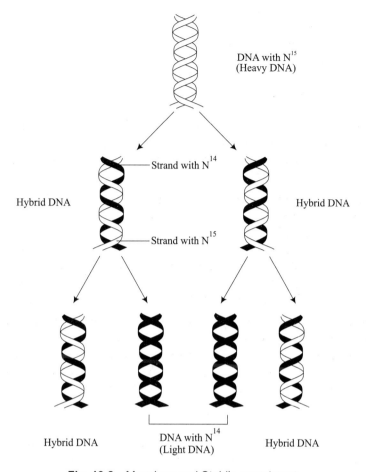

DNA with N^{15}
(Heavy DNA)

Strand with N^{14}

Hybrid DNA

Hybrid DNA

Strand with N^{15}

Hybrid DNA

DNA with N^{14}
(Light DNA)

Hybrid DNA

Fig. 19.2 Meselson and Stahl's experiment

origins. The nucleotide sequence of origins contain A, T rich sequences. The unwinding of the parental DNA occurs with the help of specific enzymes, *helicases.* This untwisting reaction requires energy which is derived from ATP.

Once helicases have unwound DNA at an origin, another enzyme *primase* binds to the helicases and denatures DNA. The complex of the primase and helicases with DNA is called *primosome.* Primase forms a "short RNA primer" complementary to the unwound template strands. The primer is then elongated by a DNA polymerase to form a new daughter DNA strand which is complementary to the template strand.

As the double stranded DNA unwinds, two single stranded template strands are exposed and form a Y-shaped structure, called a *replication fork* (Fig. 19.3).

The two strands in DNA are antiparallel to each other. As DNA polymerases can only synthesize new DNA in the $5' \rightarrow 3'$ direction, a new DNA cannot be synthesized continuously on the $3' \rightarrow 5'$ strand.

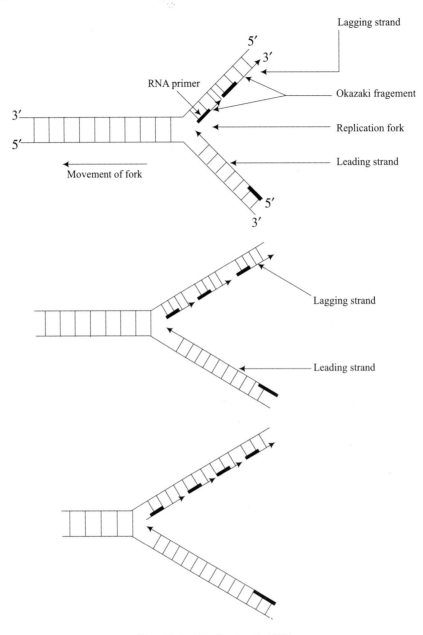

Fig. 19.3   Replication in DNA

Synthesis of one daughter strand proceeds continuously in the $5' \rightarrow 3'$ direction (in the same direction as the direction of replication fork movement). This strand is called the *leading strand*. In the other strand, which is called the *lagging strand* synthesis of new DNA occurs in the opposite direction from the movement of the replication fork. For this new primers are synthesized as more of the strand is exposed by unwinding. Each of these primers forms discontinuous DNA segments. These segments are called *Okazaki fragments*, after the name of their discoverer Reiji Okazaki. The RNA primers of Okazaki fragments are removed and replaced by DNA chain growth from neighbouring Okazaki fragments. Enzyme *DNA ligase* joins the adjacent Okazaki fragments.

Several experiments have shown that replication of DNA is bidirectional in nature i.e. DNA synthesis proceeds in both directions from the point of replication (Fig. 19.4).

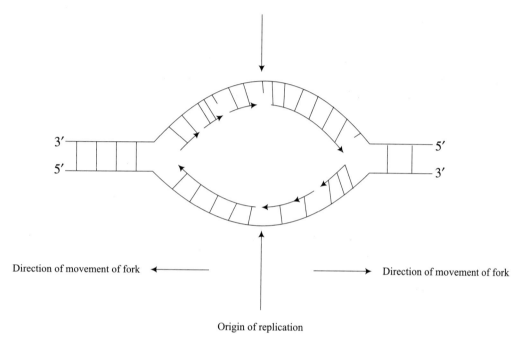

Fig. 19.4 Bidirectional DNA replication

19.2 TRANSCRIPTION

Proteins are the polymers of amino acids. Amino acids are linked together to form a polypeptide chain. The polypeptide chain forms a protein. The sequence of amino acids in a protein is ultimately determined by DNA. Thus, there is a continuous flow of information from DNA to protein via RNA.

$$\text{DNA} \longrightarrow \text{RNA} \longrightarrow \text{Protein}$$
$$\underbrace{\qquad\qquad}_{\text{Transcription}}\ \underbrace{\qquad\qquad}_{\text{Translation}}$$

This is known as the *central dogma* or one way flow of information.

Transcription is the formation of mRNA from DNA. In the process, one strand of DNA acts as a template for the synthesis of mRNA. Thus, the mRNA which is synthesized is identical in sequence with one strand of DNA and is complementary to the other strand (template). In addition to various ribonucleotides, the process requires the enzyme *RNA polymerase*.

The basic mechanism of transcription is similar in prokaryotes and eukaryotes.

19.2.1 Transcription in Prokaryotes

Transcription can be divided into three stages: (1) Initiation, (2) Elongation, and (3) Termination.

1. *Initiation*: Transcription initiates by the binding of the enzyme *RNA polymerase* at specific sequences in the DNA. These sequences are called 'promoter'. These sequences are found at −35 and −10 upstream from the base pair at which transcription starts.

   In prokaryotes, a single RNA polymerase is responsible for synthesis of different kinds of RNAs (mRNA, tRNA and rRNA). A complete RNA polymerase (Holoenzyme) consists of the core enzyme bound with a polypeptide known as sigma factor (σ). The core enzyme consists of four polypeptide chains – two α, β and β′. The enzyme can be represented by αα ββ′. The sigma factor is loosely attached with the core enzyme and is essential for the recognition of a promoter sequence. If sigma factor is not present, the core enzyme binds to DNA at various places but transcription is not initiated effectively.

   As the RNA polymerase binds with promoter, it leads to unwinding and separation of double helical DNA. One of the strand of DNA acts as a template for the synthesis of mRNA. The enzyme starts transcription at the template DNA strand. The rate at which transcription is initiated varies from gene to gene. As the enzyme moves along, the unwound region also moves with it (Fig. 19.5).

2. *Elongation*: The elongation of mRNA chain occurs with the help of RNA polymerase and ribonucleotides. Elongation is always in 5′ → 3′ direction.

   Once 8 to 9 RNA nucleotides have linked together, sigma factor dissociates from the RNA polymerase and can be used again for the initiation of other transcription reaction. The elongation of mRNA continues until the terminator sequence is reached.

3. *Termination*: Termination of the chain is signalled by specific sequences on the DNA molecule. These are called terminators. Termination of transcription of some genes in *E. coli* requires the presence of a specific protein called, *rho factor* (ρ). In other genes, core RNA polymease can itself terminate the chain at the terminators.

19.2.2 Transcription in Eukaryotes

In eukaryotes several different RNA polymerases are required for the synthesis of different types of RNAs. *RNA polymerase I* is present in the nucleus and is responsible for rRNA synthesis. *RNA polymerase II* synthesizes mRNA. *RNA polymerase III* is responsible for the synthesis of tRNA and 5S rRNA. In eukaryotes, RNA polymerase II cannot recognize the promoter alone. It requires some specific factors called *transcription factors* to bring about initiation. Different RNA polymerases require different transcription factors. These transcription factors either bind to some specific DNA sequences of the promoter or to the RNA polymerase II at the time of initiation of transcription.

In eukaryotes, transcription results in the formation of pre-mRNA (Hn RNA or heterogeneous nuclear RNA). Pre-mRNA is different from the mature mRNA in its structure. The size of pre-mRNA is much more than mRNA. Pre-mRNA undergoes certain modifications to produce a mature functional mRNA. At the 3′ end of the pre-mRNA there is addition of about 200 nucleotides of

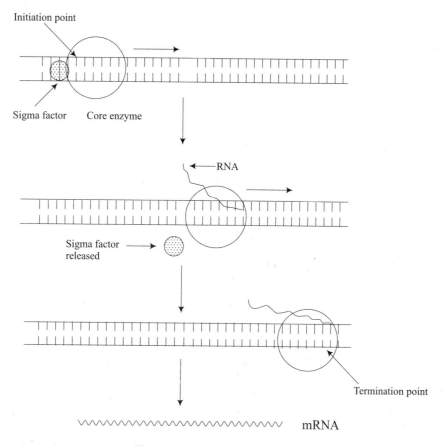

Fig. 19.5 Transcription of mRNA

adenylic acid (Poly A). The poly A tail remains as the pre-mRNA is processed to the mature mRNA. Poly A present at the 3′ and of the mRNA is generally called as 'tail'. Poly A tail provides stability to mRNA molecule. There is also an addition of a guanosine nucleotide (7-methyl guanosine) at the 5′ end. This process is called *capping*. This methylated 5′ end (cap) is essential for the ribosome to bind to the 5′ end of the mRNA at the initial step of translation. Pre-mRNA also contains non-amino acid coding sequences (introns). These introns are removed during these post transcriptional modification. The mRNA, which is formed, undergoes translation (Fig. 19.6).

19.3 REGULATION OF GENE EXPRESSION

19.3.1 *Regulation in Prokaryotes*

The regulation of protein synthesis may be either at the level of transcription or at the level of translation. On the basis their studies in *E. coli* Jacob and Monod suggested that the action of most of the genes is regulated at the level of transcription. The regulation at the transcription level may be either by *induction* or by *repression* of the gene.

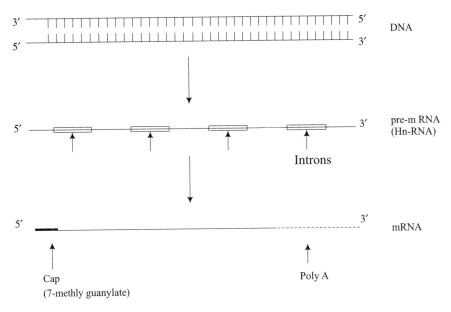

Fig. 19.6 Transcription of mRNA in eukaryotes

In induction, the gene is induced to produce an enzyme/a protein. The substance which induces the protein synthesis is called an *inducer*.

In repression, the activity of a gene is suppressed and the synthesis of a specific protein is stopped or suppressed by some substance. Such substance is called a *repressor*.

In addition to these, there are *constitutive genes* which are expressed at more or less constant rate in almost all the cells. Their expressions are not under regulation.

In 1961, Jacob and Monod proposed a hypothesis to explain the regulation of genes. This is known as *Operon model* or *Operon concept*.

An operon is a segment of DNA which consists of:

1. Structural genes
2. Regulator gene
3. Promoter gene, and
4. Operator gene.

- *Structural genes* carry the codes for the synthesis of proteins.
- *Regulator gene* transcribes an mRNA which synthesizes a protein which acts as a repressor substance.
- *Promoter gene* is situated between regulator and operator gene. It is the site at which RNA polymerase binds. It initiates the transcription of structural genes. The regulator and operator genes have their own promoter genes.
- *Operator gene* controls the expression of structural genes. The repressor substance produced by the regulator gene binds with the operator gene and controls the structural genes.

Regulator, promoter and operator all three together are called *control genes* and control the expression of structural genes.

$$\boxed{\text{Operon}} = \boxed{\text{Control genes}} + \boxed{\text{Structural genes}}$$

= Regulator + Promoter + Operator + Structural genes

19.3.1.1 Lac Operon

In case of *E. coli* a group of three genes is associated with the synthesis of three enzymes *β-galactosidase, galactose permease* and *thiogalactoside transacetylase* involved in the catabolism of lactose.

The three genes, represented as cistron Z, cistron Y and cistron A are the part of structural genes. They are closely linked and controlled in a coordinated way. These genes transcribe to form a single large mRNA with three independent translation units for the synthesis of three distinct enzymes. Such an mRNA which codes for more than one protein is called *'poly-cistronic mRNA'*. The expression of the structural genes is under the control of closely associated control genes-regulator, promoter and operator genes (Fig.19.7).

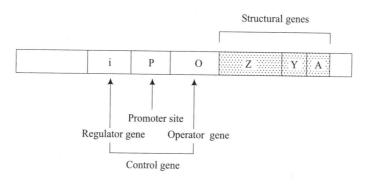

Fig. 19.7   Structure of Lac-Operon

In the absence of an inducer-lactose, regulator gene produces a protein repressor molecule which has high affinity for operator gene. It binds specifically with the operator gene. This binding of repressor molecule with operator gene prevents the binding of RNA polymerase to the promoter site and thus prevents the transcription of structural genes. The structural genes cannot synthesize the enzymes. Thus, in the absence of lactose, enzymes involved in its catabolism are not synthesized. Here, the regulator gene acts as a negative regulator of gene expression as the substance produced by regulator gene prevents the synthesis of proteins by the structural genes (Fig. 19.8).

When an inducer-lactose is added in the medium, the inducer molecules (lactose) bind with the repressor molecules and prevent their binding to the operator. The operator induces the binding of RNA polymerase on the promoter site. This results in the transcription of structural genes and synthesis of the enzymes (Fig. 19.9).

19.3.1.2 Trp Operon

In *E. coli* Trp operon consists of five structural genes and promotes the production of tryptophan in the absence of tryptophan in the environment. The five structural genes code for five polypeptide

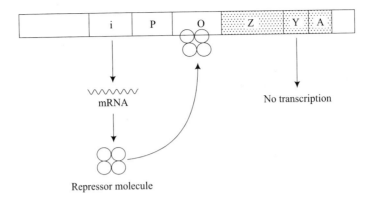

Fig. 19.8 Repression of Lac Operon in the absence of an inducer.

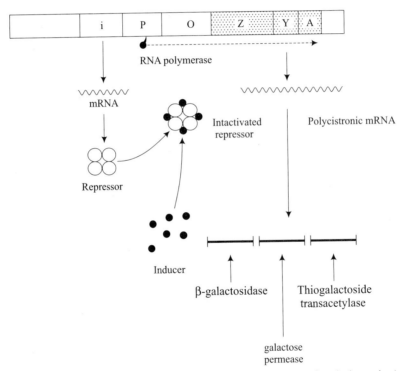

Fig. 19.9 Expression of structural genes in the presence of an inducer-lactose

chains which synthesize three enzymes-*anthranilate synthase, N-(-5′-phosphoribosyl)-anthranilate isomerase Indole-3-glycerol phosphate synthase* and *tryptophan synthase*, required for the synthesis of tryptophan from chorismate. These genes are controlled in a coordinated way. Transcription starts at the promoter site, present at the upstream of structural genes. A repressor (protein) is coded by trp R gene which is not linked to the operator. The repressor has low specificity for the operator and remains inactive. Thus, in the absence of tryptophan structural genes are transcribed to form five polypeptides which, in turn, form the three enzymes (Fig. 19.10).

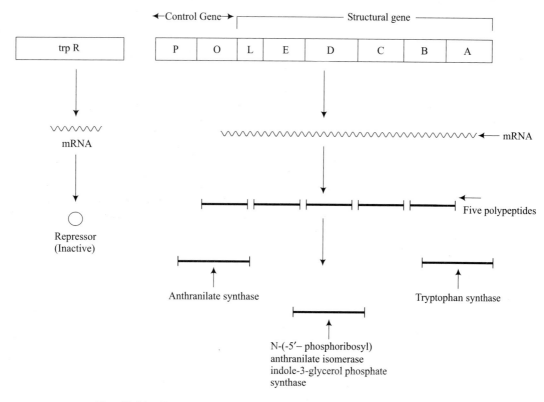

Fig. 19.10 Expression of structural genes in absence of tryptophan

When tryptophan is present, it binds with the 'repressor' and forms a complex. The repressor undergoes a conformational change and binds to the operator, preventing the transcription of structural genes. Here, tryptophan acts as a co-repressor (Fig. 19.11) Besides acting as a co-repressor, tryptophan also acts by attenuation through trpL (a leader sequence present at the upstream of structural genes).

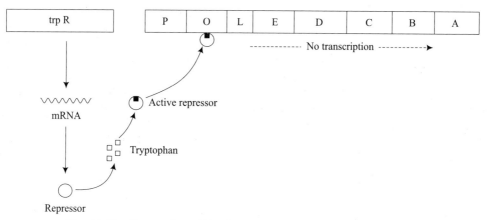

Fig. 19.11 Repression of Trp-Operon in the presence of tryptophan

19.3.2 Regulation in Eukaryotes

Regulation of gene expression is more complex in eukaryotes than prokaryotes. Since the organization of gene is more complex and the size of the genome is significantly large, more elaborate mechanisms are required in eukaryotes for gene regulation. A variety of mechanisms are employed by eukaryotic cells to regulate gene expression.

The protein coding genes of eukaryotes have large sequence of DNA which do not code for amino acids but are involved in regulation of transcription. These are called *regulatory elements*. These regulatory elements can bind to specific transcription factors and regulatory factors. They may have either a positive or a negative effect on gene expression, depending on the regulatory element.

There are certain other DNA sequences which promote the synthesis of RNA. These are known as *enhancers*. Enhancers are required for maximal transcription. *Silencers* are the other specific sequences of DNA which have opposite function. Silencer elements repress gene transcription.

DNA is tightly bound to the basic protein-histones in chromatin. Five major histones are present in the chromatin. Location (position) of the DNA in the chromatin and histone proteins themselves have some regulatory function in gene expression.

During development in eukaryotes, a phenomenon known as gene amplification is commonly observed. In gene amplification, expression of gene is increased several folds.

19.4 SUMMARY

* DNA replication takes place by semiconservative method in which separated parental strands act as templates for the synthesis of complementary daughter strands.
* Replication occurs through the action of enzyme DNA polymerase. Replication always occurs in $5' \rightarrow 3'$ direction.
* During replication leading strand is synthesized continuously while the lagging strand is synthesized discontinuously in the form of Okazaki fragments.
* Transcription is the process by which RNA is synthesized by a DNA template. The genetic information is transcribed from DNA to RNA.
* Transcription occurs in three stages-initiation, elongation and termination.
* In prokaryotes a single RNA polymerase synthesizes different kinds of RNAs.
* Sigma factor recognizes the promoter sequence and initiates the process.
* In eukaryotes, three RNA polymerases are required to synthesize different types of RNAs.
* The basic process of transcription is same in prokaryotes and eukaryotes.
* Gene expression may be controlled at the level of transcription as well as translation.
* Prokaryotic gene expression is controlled at the level of transcription.
* Regulation at the transcription level may occur either by induction or by repression of the genes.
* Regulation of gene expression is more complex in eukaryotes.

EXERCISE

1. What are different models of DNA replication? How does Meselson-Sthal's experiment support the semiconservative method of DNA replication?

2. Describe the transcription process in prokaryotes.

3. Write note on:
   (i) Trp-operon
   (ii) Gene regulation in eukaryotes.
   (iii) RNA polymerase

4. Describe the regulation of lac operon by lac repressor.

Translation: Protein Synthesis

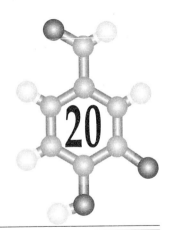

DNA is the hereditary material which stores genetic information. This genetic information is transferred from DNA to the protein. The process of transformation of information from DNA to protein occurs in two steps: First, the base sequence of DNA is transcribed into RNA. This process is called transcription. Transcription results in the formation of mRNA which is complementary to DNA.

In the second step, the base sequence of RNA is translated into the amino acid sequence of protein. This is called translation. Translation results in the formation of polypeptide (protein) (Fig. 20.1). Amino acid sequence in a polypeptide chain is determined by the base sequence of mRNA. A sequence of three nitrogenous bases in DNA/RNA specifies a particular amino acid. This is termed as *codon*.

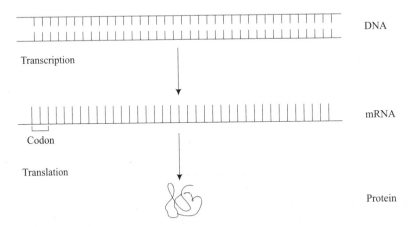

Fig. 20.1 Flow of information

20.1 GENETIC CODE

During protein synthesis, DNA forms mRNA which in turn forms protein. The type and nature of a protein depend on the types and sequence of amino acids which form that protein. The types and sequence of amino acids depend upon the nucleotide sequence of mRNA. There are four types of nucleotides in mRNA corresponding to four different nitrogenous bases. These four bases code for 20 amino acids.

If a single base codes for an amino acid (singlet codon) only four amino acids can be coded. A doublet codon would have 16 codons and would not be sufficient for 20 amino acids. A triplet codon would have 64 (4 × 4 × 4) codons and would be sufficient enough for 20 amino acids. Experiments have proved that an amino acid is coded by a group of three bases called as *codon* (Fig. 20.2).

First Letter		Second Letter								Third Letter
		U		C		A		G		
U		UUU	Phe	UCU	Ser	UAU	Tyr	UGU	Cys	U
		UUC	Phe	UCC	Ser	UAC	Tyr	UGC	Cys	C
		UUA	Leu	UCA	Ser	UAA	Stop	UGA	Stop	A
		UUG	Leu	UCG	Ser	UAG	Stop	UGG	Trp	G
C		CUU	Leu	CCU	Pro	CAU	His	CGU	Arg	U
		CUC	Leu	CCC	Pro	CAC	His	CGC	Arg	C
		CUA	Leu	CCA	Pro	CAA	Gln	CGA	Arg	A
		CUG	Leu	CCG	Pro	CAG	Gln	CGG	Arg	G
A		AUU	Ile	ACU	Thr	AAU	Asn	AGU	Ser	U
		AUC	Ile	ACC	Thr	AAC	Asn	AGC	Ser	C
		AUA	Ile	ACA	Thr	AAA	Lys	AGA	Arg	A
		AUG	Met	ACG	Thr	AAG	Lys	AGG	Arg	G
G		GUU	Val	GCU	Ala	GAU	Asp	GGU	Gly	U
		GUC	Val	GCC	Ala	GAC	Asp	GGC	Gly	C
		GUA	Val	GCA	Ala	GAA	Glu	GGA	Gly	A
		GUG	Val	GCG	Ala	GAG	Glu	GGG	Gly	G

Fig. 20.2 The Genetic Code

The first elucidation of a codon was done by Marshall Nirenberg and Heinrich J. Matthaei in 1961. H.G. Khurana, Nirenberg and Holley received the Nobel Prize for their work on genetic code in 1968.

20.1.1 Salient Features of Genetic Code

Genetic code has following features which have been confirmed on the basis of experimenal evidences:

(i) *Triplet codon*: Genetic code is triplet in nature. A codon is made up of three nitrogenous bases of mRNA arranged in a specific sequence.

(ii) *Degeneracy*: Most amino acids are coded by more than one codon. This feature of genetic code is termed as *degeneracy*. Codons that represent the same amino acid are called synonyms. Often the third base of a codon is not very significant in determining the specificity of a codon. The reduced specificity at the last position is called *wobbling phenomenon*. Only two amino acids–methionine and tryptophan have a single codon.

(iii) *Universality*: The genetic code is same in most living organisms. This phenomenon is called universality of the code. However, there are exceptions to the universality. The genetic code has been found to differ for a few codons is some ciliated protozoans, in *Acetabularia* and in mitochondrial genome. For example, in mitochondrial genome, AUA codes for methionine instead of isoleucine. UGA codes for tryptophan rather than as a stop codon. AGA and AGG which code for arginine in normal condition, act as stop codon in mitochondria.

(iv) *Non-overlapping*: There is no overlapping of codon. All codons are independent sets of 3 bases. No base functions as a common member of two different codons.

(v) *Commalessness*: In genetic code, codons are arranged in a continuous structure. No punctuations are needed between any two words i.e. there is not one or more nucleotides between consecutive codons.

(vi) *Unambiguity*: In genetic code there is no ambiguity about a particular codon. A particular codon always codes for the same amino acid and does not incorporate any unspecified amino acid into the peptide chain.

(vii) *Collinearity*: It has been well established that there is a linear correspondence in base sequence in DNA and mRNA and amino acid sequence in protein.

(viii) *Initiation and termination codon*: In most mRNAs AUG acts as the *initiating codon*. In a few bacterial mRNAs, GUG is used as the initiator codon. UAA, UAG and UGA are the *stop (termination) codons*. They do not specify amino acids and whenever they are present in mRNA, they would bring about the termination of polypeptide chain.

20.2 TRANSLATION

Translation is the process in which sequence of mRNA is translated into the sequence of amino acids of polypeptide chain. The process of translation involves the following steps:

(i) Activation of amino acids,

(ii) Initiation,

(iii) Elongation, and

(iv) Termination

(i) *Activation of amino acids*: Amino acids occur in the cytoplasm in inactive form. In this form they cannot participate in protein synthesis. Amino acids need to be activated before they can be incorporated into the polypeptide chain. Amino acids are activated by ATP which unites with amino acid and forms a highly reactive amino acid-phosphate-adenyl compound, known as *aminoacyl-adenylate*.

$$\underbrace{a.a. + ATP}_{\text{Inactive amino acid}} \longrightarrow \underbrace{AA\sim AMP}_{\text{Aminoacyl adenylate}} + PPi$$

The reaction is catalyzed by *aminoacyl tRNA synthetase*. There are specific enzymes for each amino acid. Thus there are at least 20 different *aminoacyl tRNA synthetases*. Each enzyme has double specificity. It recognizes its amino acid and also selects its tRNA. Once aminoacyl adenylate complex is formed it reacts with specific tRNA and transfers the amino acid to the tRNA. This results in the formation of *aminoacyl-tRNA complex*. This reaction is accompanied with the liberation of AMP.

$$AA\sim AMP + t\,RNA \longrightarrow \underbrace{A\sim t\,RNA}_{\text{Aminoacyl-tRNA complex}} + AMP$$

For each amino acid there is a specific tRNA. This means at least 20 different tRNAs are required for 20 amino acids.

(ii) *Initiation*: The initiation of polypeptide chain is always brought about by the amino acid methionine. The amino acid is coded by AUG, which is known as *initiation codon*. Some initiation factors are also required for the process. In prokaryotes three initiation factors–IF-1, IF-2 and IF-3 are essential for the initiation process. In eukaryotes at least 10 initiation factors are known.

In prokaryotes initiation of chain requires formylated methionine. met-tRNA undergoes formylation in the presence of enzyme *transformylase* to form *f-met-tRNA*.

$$met\text{-}tRNA \xrightarrow[\text{transformylase}]{} f\text{-met- }tRNA$$

In prokaryotes there are two types of $tRNA^{met}$ molecules. One ($tRNA^{met}$) binds with methionine and deposits it at the intercalary position in the polypeptide chain. The other $tRNA_f^{met}$ accepts formylated methionine and deposits it as the first amino acid of the polypeptide chain.

Formation of Initiation Complex:

The small unit of ribosome, 30S subunit attaches to the 5′ end of mRNA with the first codon AUG and forms Pre-initiation complex (30S-mRNA complex). Initiation factor IF-3 is required for this process. f-met-tRNA$_f$ attaches to this complex and forms 30S-mRNA-f-met tRNA$_f$. This requires IF-1, IF-2 and GTP. Now the large subunit of ribosome-50S binds with it and completes the formation of 70S ribosome. The reaction requires IF-2 and GTP.

$$30S + mRNA \longrightarrow \underbrace{30S \sim mRNA}_{\text{Pre-initation complex}}$$

$$30S \sim mRNA + f\text{-met} - tRNA_f \longrightarrow 30S - mRNA - f\text{-met} + tRNA_f$$

$$30S \sim mRNA - f\text{-met} - tRNA_f + 50S \longrightarrow \underbrace{70S - mRNA - f\text{-met} - tRNA_f}_{\text{Initation complex}}$$

In eukaryotes also the initiation of the chain occurs by methionine but formylation of methionine is not required. 40S-sub unit of ribosome binds with met-t RNA and forms the pre-initiation complex.

In eukaryotes 5′ end of mRNA has methylated 'cap.' This cap helps in the binding of mRNA with pre-initiation complex. Several initiation factors-eIF-4E, eIF-4G and eIF-4A are required for this binding. 60S subunit of ribosome binds with 40S ribosome subunit of pre-initiation complex and forms 80S initiation complex. GTP is required as a source of energy (Fig. 20.3).

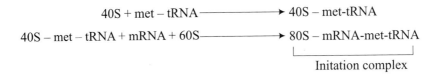

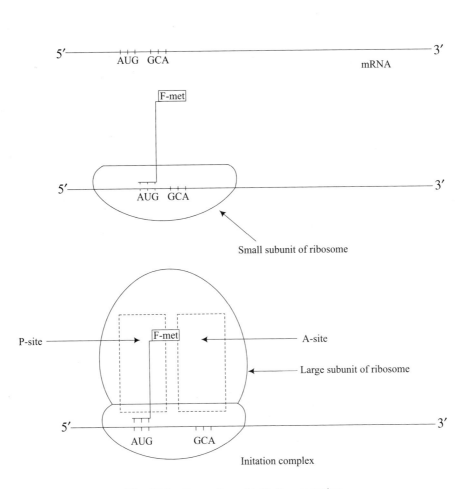

Fig. 20.3 Formation of initiation complex

(iii) *Elongation*: The large subunit of ribosome has two slots-where tRNA molecules are attached. These are: *P-site (peptidyl site)* and *A-site (amino acyl site)*. f-met-tRNA$_f$ (met-tRNA in prokaryotes) attaches to the P-site. The process requires a molecule of GTP which provides necessary energy. Another aminoacyl-tRNA attaches to the A site, corresponding the codon on mRNA. (For example if GCA is the mRNA codon, ala-tRNA will attach to the A site).

Elongation refers to the addition of amino acids one by one to the first amino acid, methionine as per the sequence of codon in mRNA (Fig. 20.4).

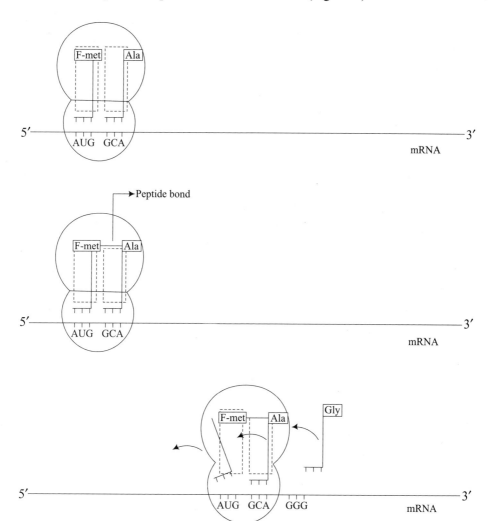

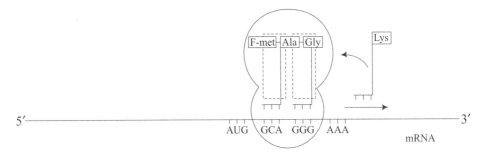

Fig. 20.4 Chain elongation during translation

When both sites (P site and A site) have amino acyl-tRNA, a peptide bond is formed between the first amino acid of P-site and second amino acid of A-site. The reaction is catalyzed by the enzyme *peptidyl transferase*, a component of 28S RNA of the 60S ribosomal subunit. After the formation of peptide bond, tRNA is released from the P-site and the polypeptide chain is transferred to tRNA present on A site.

Once the P-site becomes free from tRNA, the peptidyl tRNA shifts from A site to P-site. The process requires GTP and elongation factor EF-2. The shifting is catalyzed by the enzyme *translocase*.

As the peptidyl-tRNA shifts from A site to the P-site, A site becomes free to receive another amino acyl-tRNA as per the mRNA codon. The ribosome shifts along with mRNA in $5' \rightarrow 3'$ direction. The movement of ribosome on mRNA is called translocation and requires GTP and translocation factor EF-G.

As the ribosome move along mRNA, the initiation codon AUG becomes free to form initiation complex with smaller unit of another ribosome. As the ribosome reaches about 25th codon on mRNA, a new ribosome gets attached to the initiation codon and starts synthesizing new polypeptide chain. In this way a number of ribosomes get attached to a single mRNA. The cluster of ribosomes on mRNA is called *polyribosome complex*.

(iv) *Termination*: The process of chain elongation continues until the ribosome reaches the termination codon-UAA, UAG or UGA on mRNA. Specific releasing factors are required for recognizing specific stop codon. In *E. coli* there are three releasing factors RF1, RF2 and RF3. RF1 recognizes UAA and UAG. RF2 recognizes UAA and UGA. RF3 stimulates the termination process. In eukaryotes a single releasing factor recognizes all three stop codons. The releasing factor forms a complex with the termination codon and induces the termination of the chain.

After termination, the polypeptide chain is released from the ribosome. Ribosome dissociates into two subunits which are again used in the formation of another initiation complex.

20.3 POST-TRANSLATION PROCESSING

The polypeptide chain which is released is not functional. It undergoes certain modifications. The formyl group of the first amino acid methionine is removed by the enzyme *deformylase*. Certain other amino acids are also removed from the N-terminal or C-terminal or both ends of the polypeptide chain by enzyme *exo-peptidases*. Due to this, N-terminal amino acid methionine may also be removed. Finally the polypeptide chain singly or in association with other chains gets folded to assume tertiary or quarternary structure and becomes functional.

20.4 INTRACELLULAR TRANSPORT

After their synthesis proteins are either transported to a specific cell organelle-endoplasmic reticulum, mitochondria, chloroplast, nucleus or are secreted in the cytoplasm. The translocation of the proteins to their appropriate location is called *protein sorting* or *protein trafficking*. Proteins that are translocated to cell organelles differ from the proteins which remain free in the cytoplasm. These proteins have an N-terminal extension of about 15 to 30 amino acids called the *signal sequence*. During their transport the signal sequences are recognized by the receptors located within the membranes of the organelles.

The transport of the proteins may either be coupled with translation (co-translational transport) or proteins are transported after their complete synthesis (post-translational transport). Proteins synthesized on rough ER are translocated to the cisternal space of ER while translation is taking place (co-translational transport). Proteins destined for mitochondria, for instance, are transported after they have been completely synthesized (post-translational transport).

20.5 ROLE OF RIBOSOMES IN PROTEIN SYNTHESIS

Ribosomes are small, dense, granular particles of ribonucleoproteins. They occur freely scattered in the cytoplasm, attached to the membrane of endoplasmic reticulum (ER) and also within mitochondria and chloroplast. On the basis of their size and sedimentation coefficient, ribosomes are of two types-70S ribosomes and 80S ribosomes.

70S ribosomes are present in prokaryotic cells. The two subunits of 70S ribosome are 50S and 30S. 50S sub unit contains one molecule of 23S rRNA and one molecule of 5S rRNA and 31 proteins. 30S subunit contains one molecule of 16S rRNA and 21 proteins.

80S ribosomes are present in eukaryotic cells and consist of 60S and 40S subunits. 60S sub unit has 5S, 5.8S and 28S rRNA and about 50 different proteins. 40S subunit contains 18S rRNA and 33 different proteins.

As ribosomes play very important role in protein synthesis they are described as the 'protein factories' of the cell.

During protein synthesis, smaller subunit of ribosome (30S or 40S) binds to the first codon of mRNA and forms the initiation complex. This initiates the protein synthesis.

The large subunit (50S or 60S) has two slots-P-site (peptidyl site) and A site (amino acyl site) for the attachment of amino acyl tRNAs. Large subunit also contains enzyme *peptidase* which helps in the formation of peptide bond between the amino acid present at P site and the amino acid present at A site. This results in the elongation of peptide chain. The two subunits which get attached at the beginning of protein synthesis dissociate at the end of the process.

20.6 SUMMARY

- Nucleic acid sequence is translated into amino acid sequence in the polypeptide chain by genetic code.
- Genetic code is triplet in nature i.e. three bases code for a specific amino acid.
- Genetic code is highly degenerate and unambiguous.
- Translation is the process of transforming the information contained in the nucleotide sequence of mRNA to the corresponding amino acid sequence in a polypeptide chain.
- Activation of amino acids, initiation, elongation and termination are the four steps of translation.
- Initiation of the polypeptide chain starts at the initiation codon on mRNA.
- Ribosomal subunits assemble with aminoacyl tRNA and mRNA to form the initiation complex. Several initiation factors are required for initiation process.
- During elongation, amino acids are added one by one to the polypeptide as per the sequence of codon in mRNA.
- Termination of the chain requires specific releasing factors.
- After its synthesis polypeptide chain undergoes several modifications.
- Newly synthesized proteins are transported to specific cell organelles or are secreted in the cytoplasm.
- Ribosomes play very important role in protein synthesis.

EXERCISE

1. What is genetic code? Enumerate its characteristic features.
2. Describe the role of initiation and elongation in protein synthesis.
3. Explain the role of ribosomes in protein synthesis.
4. Write short notes on:
   (i) Activation of amino acids
   (ii) Initiation complex

Part IV

Techniques

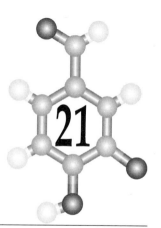

Chromatography

The separation of molecules from biological materials is an important part of biochemical work. Various methods are used for separating molecules. Chromatography is a method of separation which is utilized for separation and purification of both organic and inorganic compounds. The word 'chromatography' has its origin in Greek words – 'chromo'-meaning colour and 'graphy' – meaning to measure. The Russian Botanist Mikhail Tswett is credited with the original development of this technique. In 1903, he reported the successful separation of a mixture of plant pigments using a column of calcium carbonate. The early methods of isolation and purification of compounds of mixtures were empirical, slow and laborious. But with the advancements in separation procedures over the years, chromatography has become a highly efficient technique for separation and purification of compounds.

21.1 PRINCIPLE

Chromatography is based on *partition* or *distribution coefficient* (Kd), which describes the way in which a compound distributes itself between two immiscible phases. For two such immiscible phases A and B, the value for this coefficient is constant at a given temperature and is given as:

$$Kd = \frac{\text{Concentration in phase A}}{\text{Concentration in phase B}}$$

The term *effective distribution coefficient* is defined as the total amount of substance in one phase divided by the total amount present in the other phase.

$$\text{Effective distribution coefficient} = \frac{\text{Total amount in phase A}}{\text{Total amount in phase B}}$$

Basically all chromatographic systems consist of the *stationary phase* which may be a solid, gel, liquid or solid/liquid mixture that is immobilised, and the *mobile phase* which may be liquid or gas. This mobile phase flows over or through the stationary phase. The choice of stationary and mobile phase is made so that the compounds to be separated have different distribution coefficient.

21.2 TYPES OF CHROMATOGRAPHY

Chromatography can be classified into various types depending upon the type of solid support, stationary phase and the mobile phase. Depending upon the physical state of the stationary phase, whether the stationary phase is solid or liquid, chromatography can be divided into two types:

(i) *Adsorption*: The chromatography in which stationary phase is solid e.g. Thin layer chromatography and ion-exchange chromatography are adsorption chromatography.

(ii) *Partition chromatography*: When the stationary phase is liquid, it is called partition chromatography e.g. Paper chromatography.

Table 21.1 Various types of chromatographic techniques

Technique	Stationary phase	Mobile phase
1. Adsorption chromatography	Solid	Liquid
2. Partition chromatography	Liquid	Liquid
3. Paper chromatography	Liquid	Liquid
4. Ion-exchange chromatography	Solid	Liquid
5. Affinity chromatography	Solid	Liquid
6. Thin-layer chromatography	Solid	Liquid
7. Gas-liquid chromatography	Liquid	Gas

Chromatography can also be divided into two main groups:

(i) Column chromatography

(ii) Thin layer or planar chromatography

21.2.1 Column Chromatography

In column chromatography, the stationary phase is attached to a suitable matrix and it is packed into a glass or metal column. The mobile phase is passed through the column either by gravity or by use of a pumping system. Adsorption, ion-exchange, molecular exclusion, affinity and gas-liquid chromatographies are column chromatography.

21.2.2 Thin Layer or Planar Chromatography

Here the stationary phase, attached to a suitable matrix is coated thinly on to a glass, plastic or metal foil plate. The mobile liquid phase passes across the thin-layer plate by capillary action.

Paper chromatography is planar chromatography in which the stationary phase is supported by cellulose fibres of a paper sheet. The mobile phase passes over by capillary action over the stationary phase.

21.3 ADSORPTION CHROMATOGRAPHY

Certain solid materials have the ability to hold molecules at their surfaces. Such molecules are known as *adsorbents* and the process is known as *adsorption*. This involves weak, non-ionic attractive forces of the van der Waals and hydrogen bonding type. The binding of the molecules occurs at specific adsorption sites on the adsorbent. Various materials such as silica, alumina and activated charcol are used as adsorbents.

In this technique, column is packed with an adsorbent. When the sample is passed through the column, the molecules with specific groups bind with adsorbent. The strength of binding of a particular analyte depends upon the functional groups present in its structure. Hydroxyl groups and aromatic groups tend to increase interaction with adsorption surface. The separation depends upon the relative strength of the interaction of various molecules present in the eluent to the specific adsorption sites. As eluent is constantly passed down the column, differences in the binding strengths eventually lead to the separation of the analytes. The adsorption chromatography is influenced more by the presence of specific groups than by simple molecular size because only a specific group rather than the whole molecule can interact with the adsorption site.

Adsorption chromatography is most commonly used to separate non-ionic and water-insoluble compounds such as triglycerides, vitamins and some drugs.

21.4   ION-EXCHANGE CHROMATOGRAPHY

Ion-exchange chromatography depends on the attraction between oppositely charged particles. The technique is used for the separation of many biological materials having ionizable groups such as amino acids and proteins. Column is packed with an ion-exchanger which may be either cation or anion exchanger. *Cation exchangers* possess negatively charged groups and attract positively charged anions. These are also known as *acidic ion-exchanger* as their negative charges result from the ionization of acidic groups. *Anion-exchangers* have positively charged groups that attract negatively charged cations. These are also called *basic-ion exchange* materials since positively charges generally result from the association of protons with basic groups.

When a sample (a mixture of charged molecules such as proteins) is passed through the column carrying the ion-exchange material (e.g. cation exchanger), the molecules with greater positively charges interact with the ion-exchange material and bind tightly compared to the molecules carrying lesser number of positively charges. The molecules with negative charge may just be washed out without any binding. Then, the bound molecules are eluted out one by one with an eluting buffer of a suitable pH. A better resolution can be achieved by using a pH gradient buffer. A number of ion-exchanger resins are available which are derivatives of polymeric cellulose. The choice of the ion-exchanger depends upon the stability of the sample components, their relative molecular mass and the specific requirements of the separation (Fig. 21.1).

21.5   MOLECULAR EXCLUSION (PERMEATION) CHROMATOGRAPHY

In molecular exclusion chromatography, the molecules are separated on the basis of their molecular size and shapes. A variety of porous materials act as molecular sieve e.g. glass granules and a number of gels are used as molecular sieves. When the separation is carried out by using the gels as the molecular sieve, the technique is also called *gel filtration chromatography*. If porous glass granules are utilized as molecular sieve, the technique is called *controlled-pore glass chromatography*. The term exclusion (permeation) chromatography describes all molecular

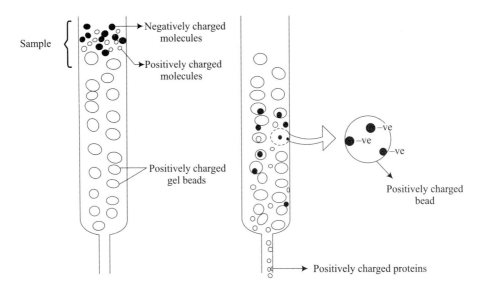

Fig. 21.1 Ion-exchange chromatography

separation processes using molecular sieves. When the mixture is poured over a column filled with gel particles or porous glass granules, large molecules in the mixture are completely excluded from the pores and pass through the interstitial spaces and appear in the effluent first. The smaller molecules are distributed between the mobile phase inside and outside the molecular sieve and then pass through the column at a slower rate, hence appear last in the effluent (Fig. 21.2).

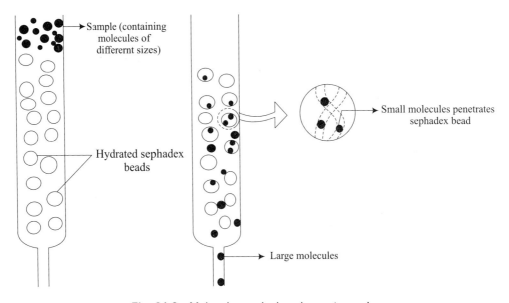

Fig. 21.2 Molecular-exclusion chromatography

21.6 AFFINITY CHROMATOGRAPHY

Affinity chromatography is the method to separate all the molecules of a particular specificity from a mixture. It is basically a type of adsorption chromatography in which the molecules to be purified are specifically and reversibly absorbed by a complementary binding substance or ligand immobilised on a insoluble support matrix. The technique is used for the separation of specific antibodies from the blood serum. For the purification of antibodies an immunoadsorbent is prepared which consists of a solid matrix to which antigen 'ag' has been coupled. The matrix used must be stable during binding of macromolecules and its subsequent elution. Agarose, sephadex and some derivatives of cellulose can be used as matrix. When the serum is passed over the immuno adsorbent, the specific antibodies in the mixture (serum) will bind with the 'ag' and retained. Other antibodies and serum proteins will pass through unimpeded. Then a reagent is passed through the column to release the bound antibodies from the immunoadsorbent. The elute is then dialyzed to remove the reagent used for elution (Fig. 21.3).

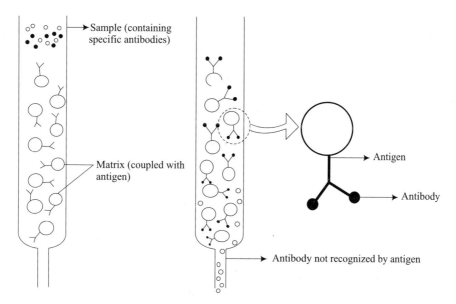

Fig. 21.3 Affinity chromatography

21.7 GAS CHROMATOGRAPHY

Gas Chromatography is a chromatographic technique which can be used to separate volatile organic compounds. The technique was introduced by Martin and Synge (1952). This technique involves injecting the volatile material into a column containing a liquid or solid stationary phase. A gas chromatograph consists of a flowing mobile phase, an injection port, a separation column containing the stationary phase and a detector. The stationary phase may be either a liquid or solid. In gas-liquid chromatography, the mobile phase is a gas and the stationary phase is a thin layer of non-volatile liquid bound to a solid support. In gas-solid chromatography a solid adsorbent acts as the stationary phase and an adsorption process occurs. The organic compounds are separated due to differences in their partitioning behaviour between the mobile gas phase and stationary phase in the column.

In this technique, the sample is injected into the injection port from where it is carried into the column. The effluent released from the column pass through a detector which is linked via an amplifier to a chart recorder which records a peak as a particular compound passes through the detector.

Gas chromatography is an efficient technique for identification of gases, pollutants, drugs vitamins, alkaloids etc. (Fig. 21.4).

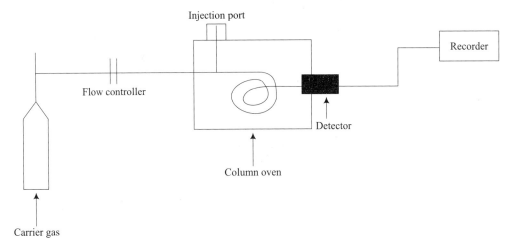

Fig. 21.4 Schematic representation of a gas chromatograph

21.8 THIN LAYER CHROMATOGRAPHY

In this technique a thin layer of stationary phase is formed on a suitable flat surface such as glass, plastic or a metal foil. The movement of the mobile phase across the layer takes place rapidly by simple capillary action. As the mobile phase moves across the layer from one edge to the opposite, it transfers analytes placed on the thin layer. The movement of the analyte is expressed by its retention factor (R_f) which is expressed as follows:

$$R_f = \frac{\text{Distance moved by the analyte from origin}}{\text{Distance moved by solvent from origin}}$$

Although the movements of compounds on TLC may be characterized by specific R_f values, these measurements are not always reproducible.

For detection of the analyte, several methods are available. A general method is to spray the plate with 50% (v/v) sulphuric acid or 25% (v/v) sulphuric acid in ethanol and heating at 110°C. This results in most of the compounds showing up as brown spots.

This technique is extensively used in quantitative determination of high molecular weight compounds. Carbohydrates, dyes, pigments, vitamins, steroids etc. can be separated and characterized by this method.

21.9 PAPER CHROMATOGRAPHY

Paper chromatography is a widely used technique in the separation of organic and biochemical products in laboratories. In this technique, stationary liquid phase is supported by cellulose fibres of a paper sheet. The mobile phase is developing solvent. The mobile phase passes over the stationary phase by capillary action. In the technique, the test solution is applied as a small spot on the chromatography paper and dried. The solvent is allowed to travel on the paper. As the solvent moves it carries the mixture components along with it. There are two types of paper chromatography: (i) Ascending, and (ii) descending chromatography. In ascending chromatography, the mobile phase (solvent) is at the bottom of the chromatography chamber. The sample spot is kept in a position just above the surface of the solvent. As the solvent moves vertically up on the paper, separation of the sample is achieved. In descending chromatography, the mobile phase (solvent) is kept in the upper position of the chromatography chamber. The paper is inserted with the upper end with sample spot close to the mobile phase kept at the upper position of chromatographic chamber. Separation of the sample occurs as the solvent moves downwards due to capillary action and gravity.

A number of methods are available for the detection of analyte. Spraying with aqueous solution of 0.2% ninhydrin saturated with butyl alcohol is commonly employed for identification. The identification of a given compound is made on the basis of its R_f value (Fig. 21.5).

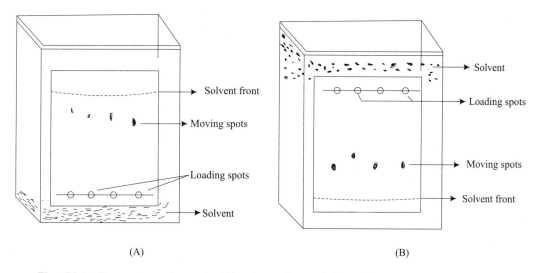

(A)   (B)

Fig. 21.5 Paper chromatography (**A**) = Ascending technique (**B**) = Descending technique

21.10 SUMMARY

- Chromatography is one of the important techniques for the separation and purification of molecules.
- Chromatography is based on the partition of solute between two phases/solvents.
- Depending upon the type of solid support, stationary phase and mobile phase, chromatography can be classified into various types.
- TLC, Ion-exchange, molecular-exclusion, paper chromatography are the different types of chromatography widely used for the separation of molecules.

EXERCISE

1. What is chromatography? Describe various types of chromatography.
2. Discuss the principle and applications of chromatography.
3. What is TLC? Describe the process of TLC.
4. Write short notes on:
    (i) Gas chromatography
    (ii) Paper chromatography

Spectrophotometry

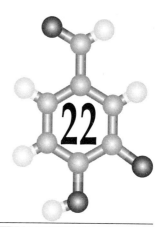

Spectrophotometry is one of the oldest methods used for quantitative analysis. It is currently used extensively in all routine analytical labs. When light is incident upon a substance, a part of it is emitted and a part is absorbed by the substance. Many instruments are used for measuring the emission or absorption of radiant energy from substances such as photometer, colorimeter and spectrophotometer. Spectrophotometer is a much sophisticated instrument. Spectrophotometry is mainly concerned with the ultraviolet and visible region of the spectrum.

22.1 PRINCIPLE

Spectrophotometer is based on *Beer-Lambert law,* which is a combination of two laws, each dealing with absorption of light by a substance.

"When light is incident upon a homogenous medium, a part of the incident light is reflected, a part is absorbed by the medium and the remainder is transmitted as such (Fig. 22.1). If I_0 denotes the incident light, I_r the reflected light and I_a the absorbed light and I_t the transmitted light, then

$$I_0 = I_r + I_a + I_t$$

If a comparison cell is used, the value of I_r can be eliminated (as it is very small, about 4%) for air-glass interfaces. Under this situation,

$$I_0 = I_a + I_t$$

The absorption of light with the thickness of medium was investigated by *Lambert.* As per Lambert's law, "When monochromatic light passes through a transparent medium, intensity of the emitted light decreases exponentially as the thickness of the absorbing medium increases arithmetically",

$$I_t = I_0 \, e^{-kt}$$

where, I_t denotes the intensity of the transmitted light, I_0 denotes the intensity of incident light falling upon an absorbing medium of thickness, 't', and k is a constant for wavelength and absorbing medium used. On changing equation from logarithms, we get,

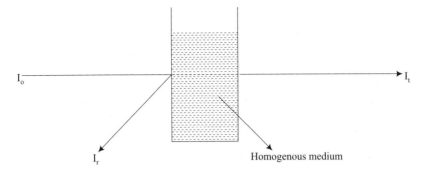

I_o = Incident light
I_r = Reflected light
I_t = Transmitted light

Fig. 22.1 Transmission of light through a homogenous medium

$$I_t = I_o . \, 10^{-0.4343kt}$$
$$= I_o . \, 10^{-Kt}$$
$$I_t/I_o = 10^{-Kt} \quad \text{or} \quad Kt = 1 \text{ or } K \propto 1/t$$

Where K is the *absorption coefficient*. It is defined as "the reciprocal of the thickness which is required to reduce the light to 1/10 of its intensity". The ratio I_t/I_o is the fraction of incident light transmitted by the medium of thickness t is termed as the *transmittance* 'T'. Its reciprocal I_o/I_t is the *opacity*. The *absorbance* (A) of the medium is,

$$A = \log \frac{I_o}{I_t}$$

Formerly, absorbance was also termed as *optical density* (OD).

Beer's law: Beer observed a relationship between transmittance and the concentration of a solution. According to this law, "the intensity of a beam of monochromatic light decreases exponentially as the concentration of the absorbing substances increases arithmetically."

$$I_t = I_o e^{-k'c}$$
$$= I_o \, 10^{-0.4343 \, k'c}$$
$$= I_o \, 10^{K'c}$$

where c is the concentration of the absrobing substance and k' and K' are the constants.

On combining both equations of Lambert and Beer, we get

$$I_t = I_o \, 10^{-act}$$

or
$$\log I_o/I_t = act$$

This is termed as mathematical statement of *Beer-Lambert law* and is the fundamental equation of colorimetry and spectrophotometry. The value of 'a' depends upon the unit of concentration. If 'c' is expressed in mole dm^{-3} and 't' in centimeters, then a is replaced by the symbol 'ε' and is termed as 'molar absorption coefficient' i.e.

$$\log I_o/I_t = \varepsilon ct = A$$

Thus, absorbance is proportional to both concentration of absorber and thickness of the layer.

22.2 INSTRUMENTATION

A Spectrophotometer has the following essential components:

    (a) Source of light   (b) Filter or monochromator

    (c) Cuvette or sample cell   (d) Photosensitive detector

The tungsten filament lamp is the most common source of visible radiation. It is the most common source of visible radiation. It is widely used in the wave length range of visible light. If colorimetric analysis is carried out in UV region of the spectrum, hydrogen lamp is used. The light source generally emits a continuous spectra. To select a narrow band from wave lengths of the continuous spectra, filter or monochromator is used. A filter is used in colorimeter. It allows light of required wavelength to pass and absorbs light of other wavelengths wholly or partially. A particular filter can be used for a specific analysis. In spectrophotometer a monochromator is used instead of a filter. A monochromator allows a large wavelength region to be scanned. A monochromator consists of an entrance slit, a dispersing element (a prism or a grating) and an exit slit. A prims splits the multi wavelength source radiation into its component parts by refraction while a grating is based on diffraction. The resolution of wavelengths is greater from gratings than from prisms. The entrace slit provides a narrow source of light so that there should be no overlapping of monochromatic images. The exit slit selects a narrow band of dispersed spectrum for observation by the detector. The sample is hold in the sample holder lying between monochromator and detector. Sample tubes are made up of glass or plastic. The sample tubes may be rectangular or cylindrical in shape.

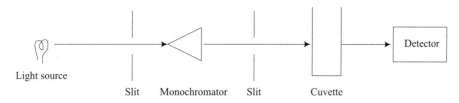

Fig. 22.2 Schematic representation of single beam spectrophotometer

A photosensitive device is used to detect the radiation transmitted from the sample. Various types of photosensitive devices such as photovoltaic cells, phototubes or photomultiplier tubes are used. Phototubes and photomultiplier tubes are much more sensitive than photovoltaic cells.

The two common forms of spectrophotometers are. A single beam spectrophotometer and a double beam spectrophotometer (Fig. 22.2, 22.3). In the double beam spectrophotometer the light beam is split into two parts, one passing through the blank or reference and the other passing through the sample at the same time. In the single beam spectrophotometer as there is a single beam of light, first the blank or reference is taken, adjustments are made and reading is taken. Blank and sample cannot be taken at the same time. In double beam spectrophotometer, as the blank and samples are taken at the same time, error due to variation in the intensity of the source and fluctuation in the detector is minimized.

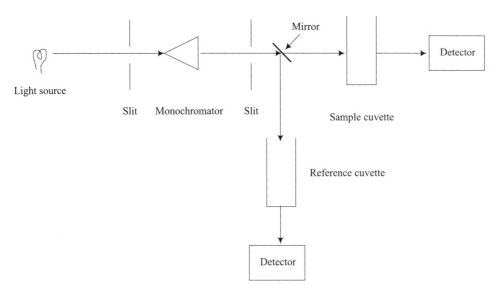

Fig. 22.3 Schematic representation of double beam spectrophotometer

22.3 APPLICATIONS

Spectrophotometry is one of the most valuable analytical techniques. It has a wide variety of applications. It provides the basis for many routine methods of analysis. Unknown compounds may be identified by their characteristic absorption spectra in the UV, visible or infrared regions of electromagnetic spectrum. The technique is widely used as a routine method for quantitative analysis of compounds.

Concentration of unknown compounds in a solution can be determined by measuring the absorption of light. Spectrophotometry is of much use in elucidating the structure of various organic molecules. Since compounds of similar structure have analogus absorption spectra, thus by comparing different absorption spectra, the structures of organic molecules can be elucidated. The technique can distinguish between 'cis' and 'trans' isomers of a complex. The geometrical isomers of compound have different visible spectra. Enzyme catalyzed reactions can be studied by measuring the appearance of product or disappearance of substrate with the help of spectrophotometry.

22.4 SUMMARY

- Spectrophotometry is an extensively used technique in biochemical labs for routine analysis of compounds.
- Spectrophotometer is based on the principle of Beer-Lambert law.
- A spectrophotometer has a source of light, monochromator, cuvette, photosensitive detector as its essential components.
- Spectrophotometry has a wide variety of applications in biochemistry.

EXERCISE

1. What is spectrophotometry? Describe the structure of a spectrophotometer.
2. Discuss the principle of spectrophotometry.
3. Write notes on:
    (i) Monochromator
    (ii) Applications of spectrophotometry

Electrophoresis

A number of techniques are used to separate molecules in biochemistry. Electrophoresis is a biochemical technique used to separate charged particles from a mixture. The term 'electrophoresis' describes the migration of a charged particle under the influence of an electric field. This is a simple, rapid, sensitive and versatile analytical tool used to study and purify the charged molecules such as amino acids, peptides, proteins, nucleotides and nucleic acids. These molecules exist in solution as electrically charged species either as cations (–) or anions (+). Under the influence of an electric current these particles will migrate either to the cathode or anode, depending on the nature of their net charge.

23.1   PRINCIPLE

When a potential difference (voltage) is applied across the electrodes, it generates a potential gradient (E). This potential gradient depends upon the applied voltage (v) and the distance between the electrodes. When the potential gradient is applied, the force on a molecule bearing a charge of q columbs is Eq newtons. It is this force which drives a charged particle towards an electrode. But the frictional force retards the movement of the charged particle. Thus, the velocity of a charged particle in an electric field is,

$$v = \frac{Eq}{f}$$

where f is the frictional coefficient.

Commonly the term electrophoretic mobility is used which is the ratio of the velocity of ion to the field strength.

$$\text{Electrophoretic mobility } (\mu) = \frac{v}{E}$$

Thus, when a potential difference is applied, molecules with different overall charges begin to move owing to their different electrophoretic mobilities. Molecules with similar charges also begin to separate if they have different molecular sizes, as they experience different frictional force.

23.2   INSTRUMENTATION

The electrophoresis equipment consists of two units – a power pack and an electrophoresis unit. Electrophoresis units are of two types-vertical and horizontal. Vertical slab gel units are commercially available and routinely used to separate proteins in acrylamide gel. The gel is formed between two glass plates that are clamped together.

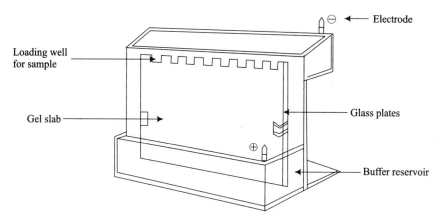

Fig. 23.1   A vertical gel electrophoresis appratus

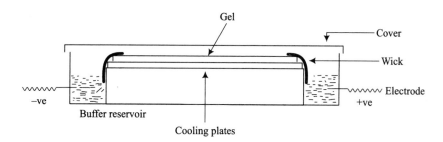

Fig. 23.2   A typical horizontal electrophoresis appratus

A plastic comb is placed in the gel solution and is removed after polymerisation to provide loading wells for samples. When the apparatus is assembled, the lower electrophoresis tank buffer surrounds the gel plates and affords some cooling of the gel plates. In horizontal gel system, the gel is cast on a glass or plastic sheet and placed on a cooling plate. The power pack supplies a direct current between the electrodes in electrophoresis unit. All electrophoresis is carried out in an appropriate buffer. The buffer is essential to maintain a constant state of ionization of the molecules being separated. The sample moves under the influence of current. At the end of the run, the position of the molecules on the gel is fixed with a fixative to prevent simple diffusion. Then the separated components are stained to visualize them. The bands can be quantitated (by elution or by scanning with a densitometer) as the uptake of the dye is directly proportional to the concentration of the molecule in each band.

23.3 TYPES OF ELECTROPHORESIS

Depending upon whether the separation is carried out in the absence or presence of a supporting or stabilizing medium, there are two types of electrophoresis. When the separation is carried out in the absence of a stabilizing medium, the method is known as *Free Solution method*. If the electrophoresis is carried out in the presence of a stabilizing medium, the technique is called *Electrochromatography* or *Zone Electrophoresis*.

23.3.1 Free Solution Method

Free solution method was first proposed by *Picton* and *Linder* (1892) but was not fully developed. *Tiselius* perfected this method and described the apparatus and methodology for which he was awarded Noble Prize. In this method, the sample solution is introduced as a band at the bottom of a U-tube which has been filled with an unstabilized buffer solution. An electric field is applied by means of electrodes, at the end of the tube. The differential movement of the charged particles towards one or the other electrode is then observed. Separation takes place as a result of differences in mobilities. The mobility of a particle is approximately proportional to its charge to mass ratio. But in this method there is always a tendency of the separated components to mix by convection as a consequence of thermal and density gradients as well as mechanical vibrations. The careful thermal regulation and isolation from mechanical vibration is difficult. Thus, free solution apparatus tends to be expensive.

23.3.2 Supporting or Stabilising Method

To overcome the adverse effects of free solution method, the electrophoresis can be carried out on a porous mechanical support. The support medium cuts down convection currents and diffusion so that the separated components remain as sharp zones. A number of supporting media are used for electrophoresis. Depending upon the type of supporting medium, there are various types of electrophoresis. Some of the common types are:

23.3.2.1 Paper Electrophoresis

Paper electrophoresis is the most common type of electrophoresis run in clinical laboratories. This technique is used for separation of small molecules such as amino acids, peptides and carbohydrates. While this technique is fine for resolving small molecules, the separation of macromolecules such as proteins and nucleic acids on such support is poor.

23.3.2.2 Agarose Gel Electrophoresis

For analysing macromolecules gel is widely used as a support medium. Starch, agarose and polyacrylamide gels are used for electrophoretic techniques. Agarose is a linear polysaccharide made up of the basic repeat unit agarobiose. Agarose gel is formed by boiling dry agarose in aqueous buffer until a clear solution forms. This is allowed to cool at room temperature to form a rigid gel. The pore size in the gel depends upon the initial concentration of agarose. If the concentration is low large pore sizes are formed. Agarose is usually used at concentration of between 1% and 3%. Agarose gels are used for electrophoresis of both proteins and nucleic acids. Agarose gels electrophoresis, run by placing the gel under the level of buffer is most commonly employed in all the nucleic acid research applications such as DNA finger printing, sequencing and recombinant DNA technology. Agarose gel is also used in techniques such as immuno-electrophoresis.

23.3.2.3 SDS-page

SDS-polyacrylamide gel electrophoresis is a widely used technique for the qualitative analysis of protein mixtures. The relative molecular mass of proteins can also be determined by this method. Sodium dodecyl sulphate (SDS) is an anionic detergent. In this method the protein mixture (sample) is first boiled in a buffer containing β-mercaptoethanol and SDS. This denatures the proteins and SDS binds with the proteins. The sample buffer also contains an ionisable tracking dye which allows the electrophoretic run to be monitored. It also contains sucrose or glycerol. This allows the sample to settle easily through the electrophoresis buffer to the bottom when injected into the loading wells. Once the samples are all loaded, a current is passed through the gel. Normally 15% polyacrylamide gel is used as the separating gel for separating proteins in the range of molecular mass of 100000 to 10000. This method can also be used to determine the molecular mass of the proteins. The molecular mass of a protein can be determined by comparing its mobility with those of a number of standard proteins of known molecular mass that run on the same gel.

Though SDS-PAGE is the most widely used technique for studying proteins but the method cannot be used for detecting a particular protein on the basis of its biological activity since the protein such as enzyme is denatured by SDS-PAGE technique.

23.3.2.4 Isoelectric Focussing

In this technique proteins are separated according to their isoelectric points. Isoeletric point is the pH at which a protein is electrically neutral and does not move in the electric field. A stable pH gradient is established in the gel by addition of appropriate ampholytes. The protein mixture is placed in a well on the gel. When an electric field is applied proteins enter the gel and migrate until reaches a pH equivalent to their isoelectric points. The focussing of the proteins to their isoelectric points makes the bands very sharp. This is a very powerful technique and is capable of resolving proteins that differ in their isoelectric points by as little as 0.01 units.

23.3.2.5 Two-dimensional Electrophoresis

When isoelectric focussing is combined with SDS electrophoresis it leads to a more sensitive analytical method than either isoelectric focussing or SDS electrophoresis alone. Two dimensional electrophoresis separates proteins of same molecular weight that differ in isoelectric points or the proteins with similar isoelectric values but different molecular weights. In this method proteins are first separated by isoelectric focussing, then the gel is laid horizontally on a second gel and the proteins are separated by SDS-PAGE.

23.4 SUMMARY

- Electrophoresis is a method of separation of particles on the basis of their electric charge.
- Electrophoresis is widely used for separation of proteins, nucleic acids, amino acids, vitamins and other such molecules.
- The technique is based on the migration of charged particles either to cathode or anode, depending on their net charge.
- The two main types of electrophoresis are – free solution method and electrochromatography or zone electrophoresis.
- In free solution method, separation is carried out in the absence of any supporting medium.
- In zone electrophoresis, some stabilizing medium is used for separation method. The stabilizing medium may be paper, starch, agarose or SDS-polyacrylamide gel, depending on the molecules to be separated.

EXERCISE

1. What is electrophoresis? Explain the principle of electrophoresis.
2. Discuss free solution method. What are its disadvantages?
3. What do you understand by zone electrophoresis? Discus the various types of zone electrophoresis.
4. Write notes on:
    (i) SDS-PAGE
    (ii) Isoelectric focussing

Glossary

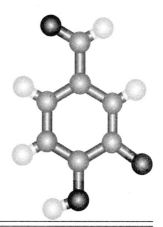

Acetyl salicylate: Commonly known as aspirin, it is an analgesic, antipyretic and anti–inflammatory agent. It inhibits enzyme cyclooxygenase required for the synthesis of prostaglandins and thromoboxanes.

Acid: A compound which can donate a proton.

Active site: Specific region of an enzyme that binds with a substrate molecule.

Affinity chromatography: A type of chromatography in which a molecule is separated from a mixture by its ability to bind specifically to an immobilized ligand.

Alodse: A sugar in which the carboxyl group is an aldehyde.

Alkaloids: Nitrogen containing organic compounds of plant origin.

Alkaptonuria: An inborn metabolic disorder which results due to the deficiency of enzyme *homo gentisate oxidase*.

Alleles: Alternative forms of a gene located on the corresponding positions on homologous chromosomes.

Allosteric enzymes: The enzymes whose catalytic activity is modulated by the binding of a specific metabolite at a site other than the catalytic site.

Amino acids: Building blocks of a protein. An amino acid consists of a carbon atom to which are attached a primary amino group, a carboxylic acid group, a side chain group, and an H atom.

Amphibolic: A metabolic process than can be either catabolic or anabolic.

Anabolism: A process where complex molecules are synthesized from simple molecules.

Anion: A negatively charged ion.

Anticodon: The sequence of three nucleotides in tRNA which is complementary to an mRNA codon.

Apoenzyme: The protein part of an enzyme (holoenzyme) which is inactive in the absence of a cofactor.

ATP: A major form of energy. It consists of adenine, ribose and a triphosphate group.

ATPase: An enzyme catalyzing hydrolysis of ATP.

Autolysis: A process in which a molecule catalyzes its own degradation.

Base: A compound which can accept a proton.

Base pair: Association between nucleic acid bases by specific hydrogen bonding. Adenine pairs with thymine and uracil or guanine pairs with cytosine.

Beri–beri: A disease caused by the deficiency of thiamine (Vitamin B_1).

β–oxidation: Oxidative degradation of fatty acids in which fatty acids are progressively degraded by successive oxidation at the β carbon atom.

β–sheet: A secondary structure in proteins which is created by hydrogen bonding between different polypeptide chains.

Buffer: Buffer is a solution of a weak acid and its conjugate base which resists any change in pH by addition of a small amount of an acid or a base.

Carbohydrates: Carbohydrates, the major source of energy, are the polyhydroxy aldehydes or ketones.

Catabolism: The metabolic process in which complex molecules are degraded into simple molecules usually accompanied by the release of energy.

Catalyst: A substance which increases the rate of a chemical reaction without undergoing a permanent change.

Cation: A positively charged ion.

Central dogma: Way of flow of genetic information in organisms. According to this, the flow of information is from DNA to RNA to proteins.

Chargaff rule: In DNA, number of adenine is equal to thymine and number of guanine is equal to cytosine.

Chromatin: A complex of DNA, histones and non–histone proteins which forms eukaryotic chromosomes.

Chromatography: A method of separation of compounds on the basis of their size, shape or charge.

Chromosome: The structural unit of genetic material consisting of DNA and associated proteins.

Chyclomicrons: Lipoprotein complexes which carry dietary lipids from the lymph through the blood stream.

Cistron: Sequence of DNA which codes for a polypeptide chain.

Citric Acid Cycle: A series of reactions which result in the oxidation of pyruvic acid into CO_2 and H_2O. It is also known as Krebs cycle or tricarboxylic acid cycle (TCA cycle).

Codon: Codon refers to a sequence of three nucleotides in an mRNA that specifies the incorporation of a specific amino acid into a protein.

Coenzyme: Coenzyme is a small organic molecule required for the activity of a complex enzyme (holoenzyme).

Cofactor: A small organic or inorganic molecule required for the activity of an enzyme.

Collagen: A group of fibrous proteins which consists of extensively cross linked molecules of tropo–collagen.

Colligative property: A physical property which depends on the concentration of a dissolved substance rather than its chemical nature.

Competitive inhibition: A type of enzyme inhibition in which a substance competes with the substrate for binding to the active site of an enzyme. In competitive inhibition substrate and inhibitor are similar in structure.

Configuration: Arrangements in space of the different groups around an asymmetric carbon atom.

Conformation: The three dimensional structure of a molecule.

Conjugate acid: A compound formed when a base accepts a proton.

Conjugate base: A compound formed when an acid donates a proton.

Co-repressor: A substance which acts with a repressor protein to block gene expression.

Cytochromes: Commonly known as electron transferring proteins, cytochromes are a group of proteins containing a heme prosthetic group.

Deamination: Hydrolytic removal of an amino group from an amino acid.

Degenerate code: When more than one codon code for an amino acid.

Dehydrogenation: Removal of hydrogen.

Denaturation: The unfolding of the native conformation of a protein by its exposure to heat or chemicals.

Dextrorotatory: An isomer which rotates plane polarized light to the right.

Diffusion: Movement of molecules from higher concentration to lower concentration.

Disaccharide: A carbohydrate composed of two monosaccharides covalently joined by a glycosidic bond.

Disulfide bond: A covalent link formed by a cystine residue.

DNA: A polymer of deoyribose nucleotides, DNA is the hereditary material of the organisms.

Eicosanoids: C_{20} compounds derived from fatty acid arachidonic acid.

Electrophoresis: A method of separation of molecules on the basis of their electric charges.

Enantiomers: Isomers which are non super imposable mirror images of each other.

Endergonic reaction: A chemical reaction which requires energy.

Enhancer: A regulatory sequence in eukaryotic DNA which modulates the rate of transcription of the associated gene.

Enzyme: An enzyme is a biocatalyst which increases the rate of a reaction.

Epimers: The stereoisomers which differ by configuration at one carbon atom.

Essential amino acids: Amino acids which are not synthesized by the body. These must be supplied in diet.

Exergonic reaction: A chemical reaction in which energy is released.

Exon: A segment of gene that codes for a part of protein sequence.

Fat: Lipid which is solid at room temperature.

Fatty acid: A carboxylic acid with a long chain hydrocarbon side group.

Feed back inhibition: When the product of a reaction sequence inhibits an earlier step it is called feed back inhibition.

Fermentation: An anaerobic catalytic process which converts carbohydrate to ethyl alcohol with the release of energy.

Functional group: A part of molecule which participates in interaction with other substances.

Gene: Sequence of DNA which encodes a polypeptide or RNA. Gene is the hereditary unit of an organism.

Gene expression: Formation of a polypeptide or RNA by a gene through transcription and translation.

Gene product: RNA or protein encoded by a gene.

Genetic code: A sequence of three nucleotides which specifies an amino acid.

Genome: The total genetic constitution of an organism.

Globular proteins: Compact and spherical molecules of polypeptide chains which are soluble in water.

Glucogenic amino acids: Amino acids which can be degraded to glucose or glucogenic precursors.

Glucogenolysis: Breakdown of glycogen.

Gluconeogenesis: The synthesis of glucose from non–carbohydrate precursors.

Glycolipid: A lipid containing carbohydrate.

Glycolysis: The process of conversion of glucose into pyruvic acid.

Glycoprotein: A protein containing carbohydrate.

Haemoglobin: A respiratory pigment of animals consisting of protein-globin and an iron containing prosthetic group known as heme.

Helicase: An enzyme that unwinds DNA.

Heme: An iron containing prophyrin, heme is the prosthetic group in the oxygen transport proteins haemoglobin and myoglobin.

Hexose mono phosphate shunt: A pathway of oxidation of glucose which yields pentose phosphate and NADPH.

Histones: Basic proteins of chromosomes. They are soluble in water and dilute acids.

Holoenzyme: An enzyme consisting of protein part (apoenzyme) and a cofactor.

Homeostasis: The maintenance of a steady state.

Hormone: A substance released by an endocrine gland which have variety of functions.

Hydrolysis: Breakdown of a compound with the help of water.

Hydrophilic: Water loving. A substance which is water soluble.

Hydrophobic: Water hating. A substance soluble in non–polar solvent and insoluble in water.

Inducer: A substance which induces gene expression.

Inhibitor: A substance which inhibits the activity of an enzyme.

Initiation factor: A protein required to initiate the process of translation.

Introns: The intervening sequences of the gene which are transcribed into pre–mRNA but are excised prior to translation.

Isoelectric point: The pH at which a molecule has no net charge.

Isomerism: The phenomenon when different compounds have same molecular formula but different physical and chemical properties.

Isomers: Compounds having same molecular formula but different physical and chemical properties.

Isotope: Atoms of same element having same atomic number but different mass numbers.

Isozymes: Multiple forms of an enzyme having similar catalytic activity but different physical and chemical properties.

Ketogenesis: The process of formation of ketone bodies.

Ketogenic amino acids: Amino acids which are degraded to yield compounds which can be converted to fatty acids or ketone bodies.

Ketone bodies: Acetoacetate, B–hydroxybutyrate and acetone, alternate forms of metabolic fuel synthesized by the liver.

Ketosis: Accumulation of excess amount of ketone bodies in the blood, body fluids and other body tissues.

Km: The concentration of a substance at which the rate of an enzymatic reaction is half the maximal.

Lagging strand: The DNA strand which is synthesized as a series of discontinuous fragments known as okazaki fragments which are later joined.

Leading strand: The DNA strand which is continuously synthesized during DNA replication.

Lesch-Nyhan Syndrome: An inborn metabolic disorder which results from the deficiency of hypoxanthine-guanine phosphoribosyl transferase (HGPRT) activity.

Levorotatory: A compound rotating the plane of polarized light to the left.

Lipids: Esters of alcohols and fatly acids.

Lipoprotein: A conjugated protein containing a lipid.

Liposome: A vesicle bound by a single lipid layer.

Lyase: An enzyme which catalyzes elimination of a group to form a double bond.

Metabolism: All the enzymatic reactions occurring in the body. It includes both anabolism and catabolism.

Metalloenzyme: An enzyme which contains a tightly bound metallic ion.

Monosaccharide: A carbohydrate which cannot be further hydrolyzed to simple sugars.

mRNA: A type of RNA which carries genetic information for protein synthesis from DNA. Genetic information from DNA is transcribed to mRNA.

Myoglobin: Heme containing oxygen carrying protein of muscles.

Non-competitive inhibition: A type of enzyme inhibition in which the inhibitor does not resemble the substrate molecule. It cannot be reversed by increasing the substrate concentration.

Non-essential amino acids: Amino acids which can be synthesized in the body and hence need not to be obtained from the diet.

Nuclease: An enzyme which breaks down nucleic acid into nucleotides.

Nucleoside: A compound containing a nitrogenous base and a pentose sugar.

Nucleotide: Monomeric unit of nucleic acid containing a nitrogenous base, a sugar and a phosphate group.

Oil: Lipid which is liquid at room temperature.

Oligosaccharide: Carbohydrate containing a few monosaccharide residues.

Operon: An operon is a prokaryotic genetic unit consisting of a promoter, operator and structural genes.

Optical activity: The property of a molecule to rotate the plane of polarized light.

Osmosis: Movement of solvent from higher concentration to lower concentration across a semi permeable membrane.

Osmotic pressure: When two solutions are separated by a semi permeable membrane the pressure which must be applied to a hypertonic solution to prevent the net flow of solvent across a permeable membrane.

Oxidation: Loss of an electron by an atom.

PAGE: Poly acrylamide gel electrophoresis.

Pentose: A sugar containing five carbon atoms.

Peptidase: An enzyme which hydrolyzes a peptide bond.

Peptide: A short linear chain of amino acids linked by peptide bonds.

Peptide bond: Linkage between the α–amino group of one amino acid and α carboxylic group of another amino acid.

pH: The negative logarithm of the hydrogen ion concentration in a solution.

Plasmid: A small circular extra chromosomal DNA present in prokaryotic cells.

Polymer: A molecule consisting of a large number of small units linked together.

Polymerase: An enzyme which catalyzes the addition of a nucleotide to a polynucleotide.

Polypeptide: A polymer consisting of amino acids linked together by peptide bonds.

Polysaccharide: A carbohydrate containing multiple units of monosaccharide.

Primary structure: The amino acid sequence of a protein.

Proenzyme: An inactive precursor form of an enzyme.

Prokaryote: A unicellular organism which lacks a well defined nucleus.

Prosthetic group: A tightly bound metallic ion or organic compound which is bound to an enzyme by covalent or non-covalent forces.

Protein: A macromolecule which consists of one or more polypeptide chains.

Purines: Double ringed nitrogenous bases. Adenine and guanine are purines.

Pyrimidines: Single ringed nitrogenous bases. Cytosine, thymine and uracil are pyrimidines.

Quaternary structure: Spatial arrangement of different polypeptide chains of oligomeric proteins.

Reduction: Gain of an electron by an atom.

Regulator gene: A gene that regulates transcription by releasing a repressor which binds the operator gene.

Release factor: A protein which induces ribosome to terminate polypeptide synthesis at the termination codon.

Renaturation: The refolding of a denatured molecule by which it regain its native conformation.

Replication: The process of synthesis of identical copy of DNA.

Replication fork: The branch point in a replicating DNA at which the two strands of the parental DNA separate from each other.

Replicon: A unit of eukaryotic DNA which is replicated from one replication origin.

Repressor: A molecule which prevents the transcription of a gene.

Reverse transcription: The process of formation of DNA from RNA.

Ribosome: A cell organelle present in both prokaryotic and eukaryotic cells which help in protein synthesis.

Ribozyme: An RNA molecule which has catalytic activity.

RNA: A polymer of ribonucleotides.

rRNA: A type of RNA present in the ribosome.

Saponification: The process of formation of soap. Fats form soap with alkali.

Saturated fatty acid: A fatty acid which does not contain any double bond.

Secondary structure: Folding of contiguous segment of polypeptide chains into geometrically ordered units.

Semi-conservative replication: Method of DNA replication in which the newly synthesized DNA contains one strand from the parent molecule and one newly synthesized strand.

Sigma factor: A part of prokaryotic RNA polymerase holoenzyme which recognizes a gene's promoter site.

Silencer: A specific sequence of DNA where a repressor of transcription can bind.

Splicing: A process by which introns are removed from a pre–mRNA and exons are joined to produce mRNA.

Stereoisomers: Isomers which are non–super imposable images of each other.

Structural gene: A gene which encodes a protein.

Teleomere: The end of a linear eukaryotic chromosome.

Termination codon: A sequence of three nucleotides which causes the termination of synthesis of a polypeptide chain.

Tertiary structure: The three dimensional structure of a protein.

Transamination: Transfer of an amino group from an amino acid to a keto acid.

Transcription: The process of formation of RNA from DNA which results in the transfer of genetic information from DNA to RNA.

Translation: The process of formation of protein from RNA in accordance to the nucleotide sequence of mRNA. This results in the transfer of genetic information from mRNA to protein.

tRNA: A type of RNA which transfers amino acids to the site of protein synthesis.

Unsaturated fatty acid: A fatty acid which contains one or more double bonds.

Urea cycle: A metabolic process which results in the formation of urea.

Vitamins: Organic substances which are required for growth and development and deficiency of which causes specific diseases. They are obtained from the diet.

Zwittor ion: A molecule which bears oppositely charged groups.

Zymogen: The inactive precursor form of an enzyme.

Index

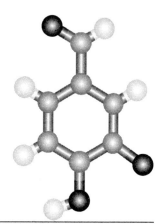